D0891831

Consuming Landscapes

Consuming Landscapes

What We See When We Drive and
Why It Matters

Thomas Zeller

JOHNS HOPKINS UNIVERSITY PRESS   BALTIMORE

© 2022 Thomas Zeller
All rights reserved. Published 2022
Printed in the United States of America on acid-free paper
9 8 7 6 5 4 3 2 1

Johns Hopkins University Press
2715 North Charles Street
Baltimore, Maryland 21218
www.press.jhu.edu

Cataloging-in-Publication Data is available from the Library of Congress.

A catalog record for this book is available from the British Library.

ISBN: 978-1-4214-4482-6 (hardcover)
ISBN: 978-1-4214-4483-3 (ebook)
ISBN: 978-1-4214-4564-9 (ebook oa)

This book is freely available in an open access edition thanks to TOME (Toward an
Open Monograph Ecosystem)—a collaboration of the Association of American
Universities, the Association of University Presses, and the Association of Research
Libraries—and the generous support of the University of Maryland. Learn more at
the TOME website, available at openmonographs.org.

*Special discounts are available for bulk purchases of this book. For more information, please
contact Special Sales at specialsales@jh.edu.*

Contents

Illustrations

Acknowledgments

Research for this book was supported by several institutions, and its writing benefited from advice and comments that I received from many colleagues. It is a pleasure to acknowledge their support.

During a fellowship at the German Historical Institute in Washington, DC, I was able to make a first foray into the comparative history of landscaped roads and to co-organize a conference on the topic. The National Science Foundation program in Science and Technology Studies supported research with grant SES-0349857; Melissa Kravetz's work as a research assistant was crucial. A John W. Kluge postdoctoral fellowship at the Library of Congress and a fellowship in Garden and Landscape Studies at Dumbarton Oaks Research Library and Collections allowed further research and writing. At the University of Maryland, the Graduate Research Board provided time for writing, and the College of Arts and Humanities funded a book workshop. The College and Dean Bonnie Thornton Dill also facilitated the publication of this book through the Dean's Covid Relief Fund Program and the Faculty Funds Competition. The University of Maryland Libraries and Dean Adriene Lim provided a generous grant to support the open access version of this publication. I am grateful to all of these institutions for the financial support that they provided.

Without the professionalism of archivists and librarians in several locations in the United States and Germany, this book would not have been possible. At the University of Maryland Libraries, Eric Lindquist, Charles Wright, Lorraine Woods, and their colleagues in the interlibrary loan department supplied books and resources from near and far. In the Department of History, Chair Philip Soergel and his predecessors Richard Price and Gary Gerstle furnished material assistance. The work of Catalina Toala, Lisa Klein, and Gail Russell, as well as their good humor, made the process so much easier.

At conferences of the American Society for Environmental History, the European Society for Environmental History, and the Society for the History of Technology, several colleagues provided formal and informal comments that helped to shape what became a book manuscript. Mark H. Rose and Bruce E. Seely deserve special thanks for sharing their extensive knowledge of American roadbuilding with me. Audiences at MIT, the University of Minnesota, the Hagley Museum and Library, and the University of Delaware gave useful feedback on various versions of this research. I am also indebted to Thomas Mierzwa (University of Maryland) for his materials on the National Scenic Byway Study of 1991, Rita Suffness (Maryland State Highway Administration) for historic preservation materials, Georg Rigele (Vienna), and Richard Hirsh (Virginia Tech) for suggestions and help with research.

During my time as a fellow at the Rachel Carson Center (RCC) in Munich, Ellen Arnold, Lawrence Culver, John Meyer, Chris Pastore, and Frank Zelko suggested ways to improve one of the book chapters. RCC directors Christof Mauch and Helmuth Trischler deserve thanks for making their center such an intellectually thriving place. Christof Mauch's collaboration with me at the German Historical Institute is a fond memory. At the RCC and afterward, longtime friend Thomas Lekan took the time to read chapter drafts and offered many helpful insights. Jean Thomson Black graciously reviewed an early draft. I am grateful to all of them. In Maryland, Audra Buck-Coleman and Jason Farman, fellow members of a 2017 campus writing group, gave encouragement at an important stage. A manuscript workshop in the summer of 2018 in College Park helped to improve a rough manuscript. I appreciate the thorough comments and suggestions by David E. Nye and Mark Cioc as well as those by on-campus colleagues Robert Friedel, Oliver Gaycken, Nicole Mogul, David Sicilia, and David Tomblin. My thanks also go to development editor Robert Kulik.

At Johns Hopkins University Press, Matt McAdam's and Adriahna Conway's professionalism ensured a productive publication process. Two outside reviewers gave valuable comments. Copy editor David Goehring's diligence prevented slips of the pen, and Caitlin Burke provided maps.

Portions of previously published papers have been incorporated into chapter three: "Staging the Driving Experience: Parkways in Germany and the United States," in *Routes, Roads and Landscapes*, edited by Mari Hvattum, Janike Kampevold Larsen, Brita Brenna, and Beate Elvebakk (Farnham: Ashgate, 2011), 125–38, and "Landschaften in Windschutzscheiben-Perspektive: Autobahnen, Parkways, Alpenstraßen," in *Landschaft quer denken: Theorien—Bilder—Formationen*, edited by Stefanie Krebs and Manfred Seifert (Leipzig:

Leipziger Universitätsverlag, 2012), 295–316. I thank the copyright holders for permissions to reproduce these materials.

My sons Tobias and Sebastian made sure that I would spend time away from the manuscript to explore the current century. Their mother, Karen Oslund, showed flexibility with scheduling research and conference trips. Friends and family offered much appreciated diversions and support.

My most important thanks go to my parents, Rosa and Rupert Zeller. I dedicate this book to their memory.

Consuming Landscapes

Introduction

Cars and Roads as Environmental Saviors

The Soviet satirists were surprised. When the writers Ilya Ilf and Evgeny Petrov toured the United States in an automobile in 1935, they encountered many novel people and sights, among them incurious hitchhikers and small towns with names like Moscow. Yet, one specific feature of the American landscape caught their eye: the amount of scenery they observed without having to leave their car. Each turn in the road "obediently opened up more vantage points on a beautiful view." The Soviet visitors realized that this was no accident and felt guided, as if they were touring through an art exhibit: "Roads like this are laid out with a specific goal: to show nature to travelers, to show it so that they don't have to scramble around on the cliffs in search of a convenient observation point, so that they can get the entire required quantity of emotions without ever leaving their automobile. In the exact same way, without ever leaving his car, the traveler can get the necessary quantity of gasoline at the gas stations that line American highways by the thousands."[1]

For these foreign observers, scenic amenities were akin to other roadside features; gas and vistas had become public amenities. Their ridicule had a point: fewer and fewer sights for automotive travelers were unplanned by the mid-1930s. Ilf and Petrov visited the United States at a time when the view from the road was the subject of learned inquiries and intense debates, as well as the result of work by laborers and bulldozers. What motorists should (and should not) see, how and when they should (and should not) see it, and why all of this mattered—these questions and the many ways to answer them delighted or disgusted both drivers and people who wrote about driving, kept professionals busy, and ultimately changed the ways in which drivers and passengers would see the world.

Eventually, one of the primary twentieth-century ways of experiencing nature for North Americans and Europeans became to drive through it. As more and more people had access to automobiles, and more and more roads made traveling easier, windshields served to frame a growing number of views. In the United States and Germany—two major car-loving and road-friendly nations—a process of turning the windshield into a picture window was already well underway when Ilf and Petrov embarked on their trip through the American countryside. In both countries, the governments built roads that featured landscapes, instructed drivers to reconnect with the countryside, and aimed at restoring the environments through which motorists moved. Such roads were to be nothing less than acts of reconciliation between nature and technology. That so many drivers and observers of driving not only accepted but encouraged the rise of the automobile as a way to rekindle their relationship with nature speaks to larger issues about technology and culture. In their eyes, the properly managed rise of a new transportation technology in the form of cars and roads would reconnect humans with nature. Previous technologies, especially railroads, had ripped such links asunder, they claimed.

Whether they were called parkways or scenic roads, or they bore specific designations to a place, these corridors of scenery enjoyed copious government funding, writers' blessings, and much visitation. Two specimens stand out. The most extensive examples in both countries that were sponsored by central governments were the Blue Ridge Parkway, running along the spine of the Appalachian Mountains in Virginia and North Carolina, and the German Alpine Road (Deutsche Alpenstraße) in Bavaria. Under dramatically different political circumstances, construction for both projects began in the mid-1930s. Both were born out of a desire to meld technology and nature into a restorative whole and to reconnect drivers with the environment. Operating a complex, mass-produced piece of machinery on four wheels, on elaborately designed roads, would immerse drivers and passengers in their scenic surroundings.

Investigating the cultural settings, politics, construction, and usage of the Blue Ridge Parkway and the German Alpine Road offers important insights into a central tension of twentieth-century modernity: how to reconcile rapid industrial development with environmental concerns. Nazi Germany and New Deal America were dramatically dissimilar in their ideals and practices regarding individual rights and inclusion; genocide, war, and terror remain the hallmarks of the former. It is all the more striking, then, to realize some similarities between both countries when it came to envisioning and planning new infrastructures for automobility. With the Blue Ridge Parkway and the German Alpine Road, planners in both countries sought to cushion industrial modernity, in the form of the automobile, and to impose a landscaped version of choreographed movement onto the mountainous environment traversed by these roads.

These plans encountered dissent from some locals, who questioned their utility, and from hikers and conservationists, who averred that non-motorized transport came closer to reconnecting humans to their environments. Such criticisms arose in both countries. The histories of these efforts appear to run parallel to each other, but they were also connected. Designers paid close attention to the environmental and roadside politics on both sides of the Atlantic. Professionals and politicians visited construction sites and completed roads in other countries, read travel accounts and scholarly publications, and imitated or rejected ideas and practices for scenic automotive restoration. Their work was referential. Ideas and knowledge continued to travel from one country to another with ease. But when drivers entered these highways or read about them, they were instructed to think of scenic roads as products of a national culture glorifying a nationally charged landscape. Studying the Blue Ridge Parkway and the German Alpine Road as parallel and connected projects, therefore, offers an opportunity to disentangle vernacular and Atlantic trends, and to move away from national notions of exceptionalism.[2]

In both the United States and Germany, driving was to become an act of restoration of the environment, and of recreation for drivers and passengers. Of these twin goals, however, only the latter survived the second half of the twentieth century. In fact, just thirty years after the initial enthusiasm, and Ilf and Petrov's visit to the United States, the idea of roads generating scenery became less and less praiseworthy for many drivers, writers, and planners. Increasingly, roads were to ensure traffic flow and safety, rather than produce vistas. By the end of the century, cars and roads became anathema for environmentalists, what with their thirst for fossil fuels and noxious by-products of pollution and noise. In simplified terms, the automobile and its attendant road infrastructures enjoyed a brief career as boons to nature, only to become environmental villains. More than a historical curiosity, this changing relationship raises important questions about humans, technology, and the environment.[3]

It is not only for reasons of historical accuracy that we need to understand how cars and roads could be seen as environmental saviors. By the late twentieth century, pollution from exhaust pipes, the urban destruction wrought by multilane highways, and the ecological habitats fragmented by rural roads had firmly established the status of the automobile as an environmental threat. Currently, the transportation sector contributes about one-quarter of all energy-related CO_2 emissions globally and is the single biggest source of greenhouse gas emissions in the United States.[4] Recapturing an extended historical moment when cars and roads could be seen as a benefit to the environment is not an effort to greenwash these technologies. Rather, it is important to understand how their roles and meanings, and the hopes and fears associated with them, have changed over time. To be sure, automobiles and highways have always been controversial. But the controversies were much more complicated (and much more interesting) than simply pitting cars and roads against the environment. While it would be absurd to deny the real dangers and risks posed by what historian John McNeill has called the "Motown cluster" of cars, roads, and their accoutrements, it would be facile to ignore the enthusiasm for and belief in the environmentally restorative potential of the early versions of this cluster, and to dismiss its manifestations as merely misguided and outdated.[5]

At the same time, many of the proponents of an environmentally benign automobility sought to reorder nature alongside efforts to reorder humanity, in ways that many of today's observers would find uncomfortable. By law or practice, not everybody was allowed to use these roads; for African Americans in the United States and Germans defined as Jews, such roads were spaces of exclusion, not of common experiences. While allegedly reconciling environment and technology, these scenic infrastructures highlighted social cleavages. Locals protested against them, often in vain. Some of the planners sought to use parkways as eugenic tools to clean up landscapes and people. Vehicles and humans had to stay in racially defined lanes to partake of the scenery. It is precisely these kinds of entanglements that are the subject of this book. In the end, the automobile and its purpose-built roads were neither the saviors that their early advocates envisioned nor the villains that their late-century critics excoriated.[6]

On a larger scale, this book demonstrates how some roads, especially during the middle third of the twentieth century, embodied the idea of scenic infrastructures. Scenery itself needs infrastructures to exist in the eye of the beholder; it cannot exist independently of technologies of access and dissemination, ranging from footpaths to coffee-table books and highways. What is more, infrastructures are not simply imposed upon nature; this book examines how they are one of the meeting places between environment and technology, and thus constitute them. Scenic infrastructures were both the imagined reconciliation among humans, technology, and the environment, and the result of complex and specific disagreements and negotiations over where, how, and at what (and whose) costs to build them.[7]

Secondly, I argue that scenic infrastructures became desirable and feasible only because of a surge of what I call roadmindedness. As chapter one will show in more detail, roadmindedness was the result of a social, cultural, and political process aimed at establishing roads as fundamental benefits. While other historians have used "airmindedness" to analyze aviation's meaning as a measure of national superiority, this book's more terrestrial focus allows for a complementary understanding of the roles that roads and highways had in building twentieth-century modernities. Soaring airplanes conjured up dormant dreams of human flight as well as feelings of terror during air raids. In contrast, road-based transportation was not new during that century, but roadminded individuals and groups transformed it dramatically.[8] Around 1900, it was hardly self-evident that planning, building, and maintaining roads, especially those only for automobiles and designed with scenery in mind, would be beneficial for society as a whole. Proponents of roadmindedness made such claims and argued that roads would bring forth social, cultural, and economic gains. It is evident to anyone today that they have succeeded. However, it is less obvious how roadmindedness became firmly implanted not just in the minds of experts and political leaders, but also cut a wide swath in the public imagination. To be sure, the commercial and professional interests of roadbuilding companies and civil engineers pushed this view. In addition, it acquired social and cultural resonance far beyond the confines of corporations and interested parties. Many others also embraced roadmindedness. The road to modernity was—as it were—a road.

Roadminded advocates elevated these infrastructures to the status of environmentally grounded artworks. Driving on them was to be an act of appreciation and rejuvenation. But unlike the art gallery that Ilf and Petrov were reminded of, scenic roads were subject to the environmental forces of weathering and overgrowth. Scenic infrastructures, while planned for a fleeting moment of visual intake, required long-term maintenance. Trees grew in undesirable locations and impeded views. Road surfaces required replacing and repaving after seasonal temperature changes and repeated use. Roads aged and changed, as did the environments of which they are a part. In this sense, this book regards humans and their changing views as historical actors alongside roads and their environments. All three played their parts.

Accordingly, the first chapter, "Roads to Nature," introduces the notion of roadmindedness, the idea that roads are worthy in and of themselves, and even more so when designed with scenery in mind. Local and regional road boosters in Europe and the United States added scenic highways to the repertoire of tourist infrastructures. While these efforts were particular, they arose from an international context of competition, imitation, and adaptation. Professional planners created a design vocabulary for these roads and presented them internationally.

"Roads to Power," the second chapter, uses the professional careers of Gilmore Clarke and Alwin Seifert, both landscape architects and prominent designers of scenic roads, as a window through which to understand how such plans could receive government funding and cultural resonance in both the United States and Germany. Rather than focusing on their individual achievements, I look at professional patterns and argue that the transformative vision of nature offered by these planners prevailed in both countries.

The planning, politics, and usage of the Blue Ridge Parkway and the German Alpine Road are the subject of chapter three, "Roads in Place." Both had been introduced as large-scale touristic infrastructures and environmental benefits, but they received funding from national governments only in the 1930s as work-creation projects. Brutal dictatorial simplicity in Germany and a more deliberate, more democratic process in the United States marked the planning phases. While located in different topographies, both roads aimed at scenic surplus by providing views from higher elevations. The inconclusive infrastructural politics of Nazi Germany left the Alpine Road in fragments by the end of the 1930s, while the National Park Service continues to operate the Blue Ridge Parkway, which was completed in the 1980s.

Yet, when roadbuilding boomed, from the 1950s onward, scenic considerations took a back seat to speed and utility. Chapter four, "Roads out of Place," analyzes the relative waning of scenic infrastructures during this period. Increasingly, roads came to express predictability and uniformity, not surprise and entertainment. Professional power struggles between civil engineers and landscape architects—and a non-expert expectation of roads as consistent and safe—contributed to these changes. In larger terms, such roads were no longer destined to amalgamate technology and the environment; they became technological corridors.

The view from the road, with its checkered history and changing aspirations, is tied to automobiles most prominently. Drivers and the people who spoke for them left records about the sensory qualities of driving. However, leaving these records does not make them the first ones to tie movement to scenery; nor does it mean that recognizing or appreciating what one sees comes naturally.

A starting point is walking and its relationship to views. In Europe's more stratified societies, walking at a leisurely pace for the purpose of observing scenery—in other words, perambulation—once a privilege of the few, landed at the feet of the emerging middle classes from the eighteenth century onward. Feudal rulers had long enjoyed non-productive, aesthetically pleasing strolls or rides in enclosed gardens and landscape parks. Their designers made sure to plan the scenery consumed on foot to be as attractive and entertaining as possible. At their most skillful, garden designers left little to chance, least of all what privileged walkers took in visually: meadows, hills, plants, and their combinations as sights. In the case of the English landscape garden, contemporaneous images might make those gardens appear static, but their experience was meant to be ambulatory. As one walked, views ahead would unfold as those behind closed. Variety was key; varying kinds and amounts of greenery concealed and revealed sights. As one writer puts it, "the garden was becoming more cinematic than pictorial."[9]

The management of views extended beyond the confines of parks and gardens. Most students of landscape design history have heard of the "ha-ha," the unseen ditch in lieu of a wall enabling a vista beyond the property's boundaries. Its main purpose was to surprise the walker chancing upon it right after having his view blocked; he would then utter an astonished "ha."[10] Meanwhile, the human effort that had gone into planning and maintaining the views was as hidden as the ditch. The majority of the people in the feudal park—gardeners and laborers—knew nature through labor by seeding, planting, trimming, weeding, and tending. The privileged stroller, however, knew landscape by seeing it in entertaining ways.[11] Scenic entertainment was tied to feudal privilege.

By the late eighteenth century, elites had literally opened up the views of those who had the time and leisure to experience them. Hunting grounds and feudal parks increasingly became accessible to commoners; the case of the Tiergarten in Berlin, the walls of which came down in 1742, is one of many. Urban residents, in particular the nascent middle classes, added strolling and viewing to their list of appropriate and respectable activities. Recreational walking, which included partaking of sights, and the creation of walking spaces—squares, sidewalks, urban parks—flourished from the late eighteenth century onward in both Europe and North America.[12]

While walking and its appreciation as a cultural activity were widespread, they played different roles, according to gender, status, class, and race. Up until the nineteenth century, most cities had been walkable because of their compact size. Urbanites conducted business and socialized on foot, unless they were privileged enough to be carried in sedan chairs or to ride in cabs. When cities industrialized and grew, commuting between one's dwelling and one's place of employment became more common. Factory workers tended to walk to work, even if it meant being on their feet for miles. Mechanized transportation in the form of trolleys and trams remained unaffordable for the daily workers' commute for several decades.[13] Laborers continued to walk to work and portions of the middle classes began to walk for leisure. As one historian puts it, "Americans who needed to walk everywhere had little incentive to create opportunities for arbitrary strolls, even if there was a set of rich, philosophical ideas available to invest their walking with meaning."[14] The same, of course, was true for Europeans. With decreasing necessity for the middle classes to walk to their jobs, walking and hiking without purpose became more attractive—and more aesthetically and culturally charged. Relative levels of affluence and concomitant access to mechanized transportation gave some people the choice not to use these technologies, but to walk instead. Public walking and hiking earned their place among respectable middle-class diversions. Many writers stressed the sensory and aesthetic enrichment gained by peregrination.[15] From Keats's travelogues of the Scottish Highlands to Thoreau's directionless but ruminative walking,

Climbing the Sunset Trail (1912), Asheville, North Carolina. These overdressed hikers in the Blue Ridge Mountains display their middle-class status via their clothing, destination, and gait. Outdoor recreation on foot was a matter of choice for them. Library of Congress, Prints and Photographs Collection LC-USZ62-71817

high-minded bipedalism became a valued middle-class pursuit exactly because it was *not* utilitarian. The middle-class walk or hike combined sociability, scenery, and slow movement.

But tensions remained. Women of any class learned to avoid certain times of day and night, and certain places, lest they be considered "streetwalkers." In the American South and beyond, walkers, especially those defined as Black, could be harassed or prosecuted for vagrancy. In contrast, the preservationist John Muir famously embarked on a thousand-mile trek from Kentucky to Florida in 1867–68 in pursuit of wilderness. Facing dangers and ruminating on the aftermath of the Civil War, Muir enjoyed the benefit of his status as a white male, which made him an oddity, at worst—but not a criminal.[16]

Disposable time, dress, and unhurried pace signaled the perambulations of middle-class hikers for recreation or edification. Gradations of wealth and status were also visible when urbanites decided not to walk at all. In New York in the mid-nineteenth century, the richest five percent owned a riding horse or carriage, and Central Park was the place to promenade their equine wealth.[17] Some of Central Park's paths were designed exactly for this purpose. Wheeled movement meant a different kind of landscape experience, one that emphasized longer looks and less attention to detail, given the relative gains in speed as compared to walking.[18]

But what did walkers and riders see, and what did they value? Designed landscapes in formerly feudal gardens and newly designed public parks were prescriptive and Romantic, generally speaking. As a response to industrialization and the growth of cities, designers sought to establish an ideal, harmonious tableau of vegetation. With little thought expended on non-human animals, such landscapes showcased a terrain of non-productive expanses of lawns punctuated by growth markers such as trees, shrubs, and more densely planted islands. For landscape architects such as Frederick Law Olmsted, these ensembles were to calm urbanites and reestablish a connection to an imagined nature. The entire park was a human creation, but its appearance bore as few traces of artifice as possible.[19]

If parks offered a complete sensual immersion into the sights and sounds of an idealized nature, panoramas and other more obviously constructed sights made visual consumption a common feature of late nineteenth-century amusements. A circular canvas seen from the inside by viewers, the panorama provided a visual narrative of a landscape, a war scene, or other noteworthy vistas, such as a Civil War battle, the Alps, London, or New York. Sometimes these canvases rotated by themselves to provide a sense of motion; sometimes the visitors created the rotation by walking. Since panoramas were housed in specially constructed buildings with targeted lighting, the aesthetic experience was meant to be comprehensive. Their design features aimed at total control of the visitors' views and an all-encompassing vision of a specific event or landscape.[20]

In addition to panoramas, observation towers provided access to views for many.[21] Such scenic infrastructures enabled views from far above with relatively little effort. Hiking associations in Germany often built them at vantage points to offer destinations and visual rewards for exertion. For those less comfortable with perspiration, railroad companies developed

North from High Rock near Pen-Mar Park (1903). This visitor to a mid-Atlantic amusement park run by a railroad company is surveying agricultural landscapes on the border between Pennsylvania and Maryland from an observation tower. The view from above was a part of the sojourn. Library of Congress, Prints and Photographs Collection LC-D4-16559

higher-altitude locations or made existing ones accessible. They were able to fill trains on weekends with the promise of scenic entertainment, supervising the entire experience, from boarding the train to the activities on site, as well as the vistas encountered.

In other words, by the late nineteenth century, views and vistas had become a controlled commodity, a destination for travelers, and the result of careful design and business ventures in the context of mass consumption. This was especially true for urbanites with the means to access them. A growing number of visitors would pay for access to these sights, whether obviously manufactured, in the case of panoramas, or less visibly reworked, in the case of rural landscapes seen from observation towers, where environmental forces and human labor were intermingled. The most frequent panoramic view of that period, however, was unintentional.

Railroads and Panoramic Travel

Passengers on the premier form of nineteenth-century transportation—the railroad—encountered a new regime of visuality. Simply put, it was a shock. They noticed it, some of them wrote about it, and the historian Wolfgang Schivelbusch memorably labels it "panoramic travel."[22] Moving in a railroad compartment provided a radically new view of the surrounding landscape: educated travelers who were used to studying the details of the landscape's foreground, and then proceeded to take into view the themes of the middle ground and the distant background, were now at a loss. Because of the movement of the train, the foreground no longer was an ensemble of identifiable landscape features. Rather, it became a single blur. Observant railroad passengers were thus forced to focus on the background, reducing their vision to large objects contained therein, such as mountaintops. Some passengers embraced this new, modern—if not industrialized—mode of viewing, while others bemoaned the passing of a slower, more detail-oriented pace. In perfectly phrased Romantic fashion, John Ruskin observed that "all travelling becomes dull in exact proportion to its rapidity."[23] The response to such boredom was another industrial, mass-produced object: entertaining magazines, newspapers, and books provided by merchants in railroad stations.

This new mode of perception and diversion depended on rapid movement: "The machine and the motion it created became integrated into his [the traveler's] perception: thus he could only see things in motion."[24] The view was industrial and outside of the control of passengers—unless a passenger could stop the train to enjoy the view. This happened rarely. An anecdote has it that the Austrian emperor Franz Joseph I stopped a train from Vienna that was traversing the Alps on its first voyage. His Majesty wanted to enjoy the view more carefully, thus reclaiming the lost foreground.[25] Needless to say, most passengers did not possess this royal scenic privilege. The view from the train, lateral and accelerated, was one with the operations of the railroad set by tracks, schedules, dispatchers, and train engineers.

Still, some trains, especially those traversing or ascending mountains, advertised the scenic qualities of their rides. In business terms, this was a niche market for railroad companies. In environmental terms, scenic trains stood for the packaging—indeed commodification—of the experience of landscapes through industrial technology and corporate design. On some trips, passengers left the railroad cars to partake of the views at prescribed stops. Glass-domed cars aimed to merge outside and inside. In the Rocky Mountains and the Alps, with their dramatic heights and, in the latter case, proximity to major population centers, such scenic ventures became commonplace by the late nineteenth century.[26] Consumed by many, such trips drew the ire of John Muir. Given the extent of cut trees and the operation of steam locomotives, Muir rebuked such operations, since "every train rolls on through dismal smoke and barbarous melancholy ruins."[27] But such views stayed on the fringes; access to scenery, however harmful it was to adjacent landscapes, became a popular commodity.

It was clear to contemporaneous observers that these new transportation regimes remade the environment. Railroad companies demanded growing amounts of commodities derived from nature. As voracious consumers of lumber and steel, the railroads drew on natural resources on a large scale. Tracks and ties, engines and railroad cars were used for construction and, in the case of steel, created demands for entirely new kinds of this metal.[28] The operation of railroads involved coal, of course, the mining of which altered subterranean environments.[29] With up to one-quarter of the annual timber production consumed by the railroads in the late nineteenth century, the nascent conservation movement feared a "timber famine" and successfully pushed for replanting.[30] Commodification was largely invisible to most travelers, unless they set foot in a steel mill or observed the results of clearcutting in a forest.

However, tracks and embankments, and tunnels and bridges, offered some of the most obvious changes to the surface of the earth. While Muir was concerned about forest clearcutting next to railroad tracks, American railroads were known as being less intrusive to the landscape—at least in comparison to Britain's elaborate railway earthworks. One historian speaks of a "minimalist infrastructure."[31] Given the relative paucity of capital, early American railroad tracks were laid more quickly and cheaply than British ones.[32] As prominent surface markers, they still left highly visible imprints.

The sights from the railroads were produced technologically. As mental images, they existed long before the trip and lingered long after they had taken place. Guidebooks, reports, and paintings were some of the most important media in producing sights and imbuing them with value. As national and international tourism grew in the nineteenth century, so did concern with these sights.[33] In addition to urban tourism, with its focus on the

High embankment, near Auburn, California (1865). Railroad tracks impose a new geometry on the landscape and result in loss of timber. This image shows an embankment built for the Central Pacific Railroad in the foothills of the Sierra Nevada in California. Library of Congress, Prints and Photographs Collection LOT 11477, no. 11

built environment, tourism outside of (and from) cities targeted scenery and landscapes. The middle-class tendency to imbue rural landscapes and life with beauty, quietude, and attraction even reached the working classes. Some of them had escaped the "idiocy of rural life" (Karl Marx) and agricultural labor for the drudgery of urban wage labor, but they still enjoyed the escape to the nearby countryside for brief jaunts on Sundays.[34] Extended trips were the privilege of the few.

Travelers were by definition in motion—and yet, the scenery they traveled to enjoy was thought to be static, immutable, permanent. Tourists needed to get to new locations, move around there, and then return. But in a less trivial sense, the journey and the sights—tourism and being mobile—were intertwined: "Landscapes are produced by movement, both of the senses and of the body," as one scholar of tourism puts it.[35]

Cars and Scenery

In contrast to railroads, early automotive promoters claimed that cars would enable them to regain control over their mobile lives. They no longer depended on train schedules and could start whenever they pleased. Early cars did not need special infrastructures to run, nor a central authority to control their movements. But they had eloquent, affluent advocates. As historians have pointed out, upper-middle-class motorists disdained railroad travel as common and subject to corporate control.[36] They could purchase the means to get away from the rabble. Self-propelled motion allowed for individual control of sights and vistas; motorists could stop and enjoy the view wherever they pleased. Here is how a writer put it in a German motoring magazine in 1905:

> Now the car has arrived and it has delivered the travelling nature lover from the dominating power of space, so that he, with his freedom of movement, can enjoy the speed of the railway and the comfort of the compartment. No one tells him road and purpose, time and departure. He can buzz along from place to place, he can relax at a beautiful, shaded spot with his fellow-travelers, and taste the delicacies from his basket; he can, if he so wishes, change his goal on the spur of the moment and does not have to pass the beauty of regions that are situated off the road as he is forced to do by the insensitive railway.[37]

The control of the view went hand in hand with control of the trip. However, temperamental gasoline-powered engines forced motorists to stop not only at points of their own choosing. As some historians have argued, the rugged unreliability of this type of motor only added to its allure for leisure-oriented, thrill-seeking early adopters.[38] Many of these motorists saw themselves as adventurous explorers encountering dusty roads, being exposed to the elements in open cars and overcoming engine troubles.[39]

Increasingly reliable and less expensive cars found more buyers after the end of World War I in America. This was a turning point for mobility, as mass production on assembly lines and the growth of disposable incomes allowed for the rapid appearance of the automobile, at least in the United States. By 1927, an astounding 55 percent of American households had access

El Tovar Point, Grand Canyon (1914). While obviously staged, this image speaks to the automotive exploration of scenic landscapes. Disregarding established tracks and trails, this motorist ventures to the rim of the Grand Canyon in Arizona. Grand Canyon National Park Museum Collection

to at least one car.[40] While wide swaths of Americans, especially those in urban centers and with less purchasing power, continued to rely on public transportation for daily needs, automobiles had become a consumer item owned by a majority. The middle classes, who dominated debates regarding mobility and its meaning, embraced cars for the most part. For rural residents, and especially for farmers in remote areas, cars were adaptable sources of motive power for many utilitarian purposes.[41] Although not commonplace as means of daily transportation, cars were a common sight by the 1920s, and a common means to experience sights for the middle classes by the 1930s.

But even if they owned an automobile, commuters overwhelmingly relied on subways, trams, and suburban trains for daily trips until the 1950s and 1960s. For urban and suburban car owners, weekend and holiday trips became the domain of the automobile. According to a famous sociological study, the automobile had become "an accepted essential of normal living" for the white residents of Muncie, Indiana, by the late 1920s, with one particularly enthusiastic mother of nine children claiming that she would "rather do without clothes than give up the car."[42] Car owners altered their leisure habits profoundly as the car entered social life and enabled more frequent outings. With relatively little effort and expense, camping with the automobile became a widely shared pastime.[43]

Taken together, such usages contributed to the rising popularity of the automobile. Privately owned and maintained automobiles were one of the most important and visible tokens of modern consumerism, indicating a general rise in wealth and disposable incomes.[44] In the American case, one booster argued that cars and scenic tourism could fight communism and help to transcend ethnic and class boundaries: "It is hard to convince Steve Popovich, or Antonio Branca, or plain John Smith that he is being ground into the dust by Capital when at will he may drive the same highways, view the same scenery and get quite as much enjoyment from his trip as the modern Midas."[45] According to this voice, automobility and scenery would serve as anti-Communist equalizers.

However, the rise of automobiles and roads, and the relative decline of public transport, were neither inevitable nor merely a product of their respective technical properties. As several historians have shown, government policy, social regulation, and cultural valuation all contributed to the competition between road and rail during the twentieth century.[46] In this respect, comparisons matter. For the entire twentieth century, the United States was the single largest market for automobiles. By 1940, Americans owned more automobiles than the rest of the world combined.[47] In contrast, the automobile was an imagined but no less important commodity to Germans for quite a while. The extant statistics tell us that, on average, car ownership was limited to one in 140 Germans in 1930 and one in six by 1965 in West Germany, with a growth in motorization during the 1930s and a dearth of privately owned cars right after World War II.[48] Cars were rare but talked about during the interwar years and received a major, if often rhetorical, push during the Nazi years. However, statistical averages can be misleading: only 27 percent of households had access to at least one automobile in 1962. During the postwar economic recovery, almost three-quarters of West Germans were probably fascinated by cars but lived in households that did not own one. Not until 1973 were West Germany's car-owning households in the majority and thus at par with those in the United States as measured in 1927.[49]

While these numbers would seem to indicate that the stories of automotive dominance in the United States and Germany were separated by almost half a century, they also occurred simultaneously and in an interconnected fashion. Germans looked to America and saw a profusion of automobiles in the 1920s; Americans looked to Germany and saw a profusion of roads in the 1930s. Whether as aspiration or as material reality, automobiles and roads had become part of the cultural vocabulary. Before cars appeared, trains had introduced the idea of velocity and mass movement, as had faster ocean liners. Spectacular new forms of movement such as aviation captured the fascination of many, even if they never flew. Modes of transportation, their concurrence and competition, contributed to a sense of technological modernity.[50]

At the same time, automobiles were part of different modes of consumption in Europe and the United States. Before the mid-1930s, cars remained out of reach for most European households, except for select professionals in urban settings. Consumption was class-based. Operating an automobile often was often seen as a display of power, wealth, and an arrogant protrusion into public space.[51] In Germany, carmakers continued to use quasi-artisanal production techniques.[52] By contrast, Fordism and the triad of assembly-line manufacturing, rising wages for workers, and cheaper, mass-produced automobiles dominated American consumption from the 1920s onward. Even though car culture remained aspirational, rather than pervasive, for many households, Americans had incorporated automobiles into their daily lives to a degree that was astonishing when compared to the experience of Germans and other Europeans.[53]

As the first automobiles appeared just before the turn of the twentieth century, movement and scenery were as closely coupled as railroad cars. Whether by perambulating in urban parks or hiking in the countryside, visiting enclosed panoramas, or traveling on regular railroads or in excursion cars, the experience of landscape had increasingly become a mobile one for travelers, especially those from the urban middle classes. Different modes of mobility provided different genres of views. Speed mattered greatly, as did ownership of movement. Walkers and hikers controlled their own movements in concert with their bodily limitations; a slower pace allowed for more detailed views. At the other end of the spectrum, mechanized transportation enabled the blurred railroad journey. Celebrated by some and rejected by others, this experience remained profoundly modern and industrial. Railroad vistas were the unintended results of considerations of throughput and managerial oversight. On the fringes of the railroad enterprise, scenic tours attracted customers who would pay for obviously constructed sites such as cities and indoor panoramas, but also to observe rural landscapes, the results of human and non-human forces.

In early motorists' travelogues, overcoming the constraints of panoramic travel was a common refrain. Initially, they shared existing roads with other users in often uneasy and sometimes conflictive ways. As the following chapters will show, the shape and meaning of roads in the automobile age were subject to debate and disagreement, delight and disdain.

1

Roads to Nature

Woodrow Wilson had a busy day. On June 7, 1916, the American president met with Louis Brandeis, the future US Supreme Court Justice who had survived a contentious hearing before the Senate a few days earlier.[1] Wilson probably followed the news from Europe as well. Yet, at suppertime, he participated remotely in a ritual usually reserved for local politicians—the dedication of a new road. At 8 p.m., the president pushed a button at the White House, thus closing "a circuit reaching across the Continent."[2] His fingers activated an electric magnet in Oregon, a weight dropped, and an American flag unfurled. The continent-spanning effort to connect president and flag via wires was part of the dedication ceremony for the Columbia River Highway, which connected Portland with the eastern interior of Oregon by hugging the shores of the Columbia River. With his effortless endorsement, Wilson upgraded a local thoroughfare to a matter of national attention.

Why would a sitting president participate in opening ceremonies for a regional road at a time when more pressing matters—World War I, for instance—were at hand? What, in other words, was so special about this highway? On a personal level, Wilson's endorsement of this road marked a departure from his earlier criticism of the automobile. While president of Princeton University, he had remarked in 1906 that "nothing has spread socialistic feeling more than the use of automobiles. To the countryman they are a picture of the arrogance of wealth with all its independence and carelessness."[3] A decade later, Wilson embraced roads as a means of connection, as "a thong between that community and the nation to which it belongs."

Facing the small, yet rapidly growing group of motorists, Wilson joined the chorus of those who saw promise in automobility and its infrastructures.[4]

For Wilson and others, roads such as the Columbia River Highway promised a new type of mobile scenic experience. It is curious to realize that one of the first cultural and environmental repercussions of the rise of automobility was an urge to create a new kind of moving scenery. The automobile was still new during the second decade of the twentieth century. Its place in society, patterns of use, and larger significance were not settled. Some of its promoters claimed that driving, especially on roads such as the Columbia River Highway, restored a connection to nature supposedly lost during the first wave of industrialization. As the alleged antidote to mechanical railroad travel, driving on this riverine road still depended on infrastructures and motorized movement, to be sure. But the display of scenery on this road was not accidental; it was one of its main effects and justifications.

The Oregon road is closely connected to Samuel Lancaster, its main designer and highway booster.[5] He was a civil engineer who wanted to use the road to embellish the scenery. In addition to guiding drivers to the beauty spots along the road to allow for viewing possibilities, Lancaster's design parameters for the highway set it apart from other winding roads.[6] Driving on it was not to be fraught with danger; it was meant to be uniform and safe. Given the low density of traffic, however, drivers could (and were expected to) stop at their own leisure and explore the total scenery, including the foreground, while parked. The *New York Times* mentioned natural features and artifacts in one breath: "Its beautiful waterfalls, wonderful rock formations, tunnels, cliffs, retaining walls and artistic bridges all tend to make this delightful thoroughfare America's most noted example of man's intelligent development of nature's creation."[7] Lancaster was especially proud of the viaduct and tunnel at Mitchell Point, which he declared "fully equal to the famous 'Axenstrasse' of Switzerland and one of the great features of the Highway."[8] Apparently, Lancaster and other local boosters felt that a modern staging of American scenery could make their country competitive, if not on a par, with the scenically and architecturally well-endowed countries of Europe. Such referential attitudes gave way to a more self-centered approach by the 1930s, when parkways achieved national prominence and gained the attention and resources of the federal government.

Lancaster, in his aim to make the scenic features of the river valley "easily

Mitchell Point Tunnel on the Columbia River Highway. Having toured Europe and its roads, road builder Samuel Lancaster emulated the Axenstraße in Switzerland for some of the features of the Columbia River Highway in Oregon. The Mitchell Point Tunnel (*above*) aims to outdo its Swiss counterpart (*opposite*) by featuring five windows instead of three. Samuel Christopher Lancaster, *The Columbia: America's Great Highway through the Cascade Mountains to the Sea*, 3rd ed. (Portland, Oregon: J. K. Gill, 1926), 124

accessible to all," claimed hyperbolically that the sights had been "partially hidden, and as far as the general public was concerned they might almost as well have been on the dark continent."[9] The reference to Africa was telling: roads such as the Columbia River Highway were quasi-colonial in attitude, audience, and aim. Their promoters often used the language of discovery and access when presenting the sights. Such roads were instrumental in organizing space. Building on previous mapping and naming, parts of the topography were then presented as picturesque to visitors. The audience, or colonizers, were affluent city dwellers from Portland or elsewhere. The aim, therefore, was not simply to present scenery, but to order and commodify it.[10] A local promoter and advocate of the road put it bluntly: "We will cash in, year after year, on our crop of scenic beauty, without depleting it in any way."[11] The view from such scenic roads was the result of design, planning, cultural politics, and local schemes. Gazing at such roads allows us to under-

Gallery on the Axenstraße (Switzerland). Library of Congress, Prints and Photographs Collection LOT 7738

stand why desires and fears came together in a particular historical moment and took the form of these particular roadways.

While the Columbia River Highway rarely receives praise in national newspapers today and Woodrow Wilson's legacy is defined by events other than his remote role in the road's inauguration, such roads still matter. They helped to bring forth the specific kind of ambulatory, scenic tourism that Ilf and Petrov were so astonished to find in the United States. While it was more extensive in this country than elsewhere, it drew on and was part of an international exchange of knowledge, expertise, and views regarding scenic infrastructures. Since they offered functional use and scenic recreation, and often did so simultaneously, these connections expressed roadmindedness in both North America and Europe.

Rather than a firmly defined set of beliefs and practices, roadmindedness was an uneven but successful process of prioritizing roads in society, culture, and politics. Its advocates changed over time and included varying professional and social groups and individuals with sometimes different agendas. They used various jargons and means of influence, but they all became roadminded and convinced others to do the same. For example, journalists and politicians who decried the dominant role of railroads in transportation saw roads, trucking, and automobiles as forces to counter corporate behemoths (in the United States) or a state-owned monopolist (in Germany). Civil engineers and landscape architects correctly sensed new professional opportunities. Aesthetically minded writers welcomed roads as places to regain sensuous connections to the landscapes surrounding transportation corridors. The roster of roadminded individuals included Thomas MacDonald, Robert Moses, Emily Post, and Lewis Mumford in the United States, the Anglo-French writer Hilaire Belloc, as well as Fritz Todt, Alwin Seifert, and Hans-Christoph Seebohm in Germany. As this book will show, some of these individuals remained steadfast champions of the cause while others (such as Lewis Mumford) were initially enthusiastic but became some of its most ardent critics. Some conservationists, especially in the early twentieth century, were roadminded, while many of their later versions abhorred the very idea. National governments lent administrative support and unprecedented monies to the process, thus ensuring its infrastructural longevity.

Despite these differences, one compatible goal united these roadminded voices: the idea that roadbuilding was economically, socially, and culturally beneficial. Roadmindedness included a set of attitudes, policies, and funding practices that promoted the planning, design, construction, and (to a lesser degree) maintenance of streets and roads. Roadmindedness was built on the simple yet powerful belief that roads had intrinsic value; perhaps the most important attribute of roadmindedness was to make their utility and cultural worth self-evident. It elevated roads. They used to be ancient, quotidian, and multipurpose spaces on which people, animals, and various vehicles moved. Instead, they became a modern, paved, reliable, all-weather, and all-season infrastructure for automobiles and trucks—or only cars, in the case of American parkways. The result of campaigns, books, newspaper articles, and lobbying was the firmly anchored and no longer questioned understanding of roads as emblems of economic growth, technological modernity, and even beauty. To be modern was to be roadminded. While contemporaneous observers did not use the term themselves, roadmindedness is useful for understanding this process.

During the middle third of the twentieth century, roadmindedness claimed a higher national priority in the United States and Germany in a particular form: scenic infrastructures, such as parkways and other scenic highways, received cultural validation, as well as administrative and financial resources from regional, state, and central governments. Such scenic infrastructures promised to leave behind the ostensibly chaotic and polluted landscape of transportation created by railroads, their corporations, and their users. An orderly, clean, and attractive new landscape was to appear.

While other scholars have examined the changing meanings of automobility and driving in various contexts, my goal is to compare the history of parkways and other scenic roads in Germany and the United States as interconnected icons of roadmindedness.[12] Rather than taking designers and promoters at their word and perpetuating the idea that such roads emerged from and reflected particular national settings and values, this book examines roads in these two countries side by side. Seen from this perspective, international entanglements and the circulation of knowledge on landscapes and roads figure much more prominently. Comparing does not mean equating, however. These comparisons bring into sharp relief the differences between a democracy and a dictatorship, and the tensions between mass-based consumerism and the appeal of national, iconic technologies. While Germany and the United States shared the same time frame for the high point of this set of attitudes and actions, their origins and specific histories differed.

Comparing these two roadminded countries, and the ways in which they arrived at roadmindedness and its manifestations, also allows for a fresh perspective on the history of speed and driving. Not all movement is inevitably geared toward going ever faster. At first glance, the history of transportation would appear to evolve from organic forms such as walking and horse-drawn coaches to increasingly rapid mechanized means of moving via trains, automobiles, and airplanes. But even in today's globalized economy, most freight moves at the relatively leisurely pace of large ships. Their size has increased much more dramatically than their speed. Supersonic air travel is all but dead.[13] Automobiles in cities are often stuck in traffic rather than moving speedily. The pace of automotive travel is regulated and supervised, even though the response to speed limits makes many drivers scofflaws. The history of parkways and scenic roads in Germany and the United States helps us understand how acceleration and deceleration of traffic were not the results of intrinsic advantages of particular technologies and modes of transportation. Traffic does not go progressively faster, nor does it slow down as a matter of course. Rather, different groups in society acting at different times with different values and goals declared certain speeds to be appropriate or useful. Many parkways and other scenic roads were designed with deceleration in mind. Purposefully winding roads forced drivers to go slow and appreciate landscapes, rather than hurry through them. Driving slower, however, was contested and not generally accepted. Speedy and slow movement are intertwined in much more interesting ways than the idea of constant acceleration would suggest.[14]

As an international movement, roadmindedness had no single point of origin, either geographically or chronologically. But several individuals, organized movements, state agencies, and other institutions pushed such views and practices much higher on political, social, and cultural agendas. In doing so, roadminded champions upheld this process on four distinct yet related levels: as an engineering and political movement, as an international movement, as a cultural and environmental idea, and as a touristic device. These levels of roadmindedness were intertwined in planning and building individual roads, but it is helpful to disentangle some of these strands in the following pages.

For roadmindedness, movements and institutions mattered. The Good Roads Movement in the United States and the Bureau of Public Roads in its early years, while primarily interested in utilitarian roads for farmers and commerce, helped to introduce the idea of constructing road networks for economic growth. This was not as self-evident as it might appear today. Given the slow start of motorization in Germany, no institutional equivalents existed there before 1920. In other words, roadmindedness had more vocal advocates at this time in the United States than in Germany. Even the Americans, however, referenced European roads and scenic infrastructures, especially in Alpine locations, as the Mitchell Point example shows vividly. Given the prominence of Switzerland and other Alpine countries for scenic infrastructures, their promotion of tourism via road was of major international consequence. Roadmindedness took on institutional force in the National Park Service (NPS) with its ambitious agenda of building scenic roads. International road conferences, an important forum for exchanging ideas, helped to create a body of cosmopolitan knowledge and a cadre of experts to push for its implementation in individual countries. Crisscrossing the Atlantic is necessary for understanding the different formations of roadmindedness and the ways in which they related to each other. In some cases, the connections were more than obvious; in others, they were purposely hidden.

Roadmindedness in Engineering and Politics

The growth of the road system in the nineteenth and twentieth centuries, while impressive on its own, owed a lot to related transport technologies. Even though roads and road transport have been omnipresent in human history, the nineteenth-century rise of railroads resulted in a demand for roads. This might seem paradoxical at first glance. However, the growing amounts of freight and numbers of passengers being moved across larger distances by the railroads meant that they had to be fed into and distributed from depots and train stations. While urban public transit and rail feeder lines picked up some of this traffic, coaches, cabs, trucks, and horse-drawn carts moved it on urban, suburban, and rural roads and streets.[15] The spectacular growth of industrial cities and trading hubs depended upon the concomitant rise of railroads and roads.

Roadmindedness exposed and contributed to tensions between urban centers and their hinterlands. Given that more and more vehicles with more and more freight and passengers traversed existing roads, questions of maintenance became contested, especially outside of cities. Often, locals were responsible for the upkeep, but out-of-town users were not. In the case of bicycles, and later automobiles, the displeasure, ire, and occasional physical violence meted out to early adopters had some of their roots in these issues.[16] Organized middle-class urban bicyclists were the first ones to call for a massive new roadbuilding program in the United States. This demand did not sit well with rural residents who abhorred the costs. The Good Roads Movement, as it was called from the 1880s onward, began with the pleas of cyclists on leisure outings, who envisioned smooth, hard roads instead of the dirt roads with seasonal problems that they encountered. By the 1890s, economic arguments based on throughput and ease of traffic began to dominate the conversation about good roads. The federal government initiated surveys and tallied costs and driving times for rural traffic. It built up engineering expertise, both for maintenance and for new construction. Instead of local and varied ways of building and maintaining roads, new federal and state experts sought to define good and acceptable roads.[17]

In the process, variety gave way to conformity. As historian Christopher Wells has noted, there was an unmistakably environmental aspect to this process: heat, cold, rain, snow, sleet, sunshine, wind, and time of day mattered less and less, as roads were to be passable at all but the most extreme times, predictable in their surface appearance, and practically useful.[18] Road-mindedness, in this engineering view, included a vision of all-weather, all-year access. In infrastructural terms, one of the reasons for the success of railroads was that their operation did not depend on good weather and the seasons to the same degree as did operating waterways. Now, such regularity came to be the goal of road traffic as well.[19]

These visions alone did not convince local politicians—but money did. Individual states in Germany expanded their arterial highway programs in the second half of the nineteenth century, often based on academic knowledge generated in France's engineering academies.[20] In the United States, the federal government induced states to upgrade existing roads by offering subsidies in exchange for uniformity of design by the twentieth century. Engineers became policymakers in the process, as the historian Bruce Seely argues. The foremost federal road planner, Thomas H. MacDonald (1881–1957), chief of the Bureau of Public Roads (BPR), succinctly noted that the organization of roads and road financing was more difficult than solving technical issues such as paving and road width.[21] Two federal programs, one in 1916 and one in 1921, enshrined a cooperative relationship between Washington and the states: the latter submitted proposals to the administration, which were evaluated according to necessity, as indicated by already existing traffic and conformity to design parameters. By the 1920s, the initial focus on rural roads to improve agriculture had given way to broader attention to urban and suburban transportation.[22] In 1922 alone, some 10,000 miles of new roads were constructed in the United States. Even more astonishing are the 420,000 miles of state roads built between 1921 and 1936, a period that Seely calls the "golden age of highway building."[23] While popular, these public policies were also the result of pressure groups at work: "Auto manufacturers, auto clubs, the trucking industry, and highway engineers came together to form in effect a single lobby for highway construction and maintenance, bearing on government at all levels," as two scholars aptly put it.[24]

Federal engineers were not simply responding to requests from the states. They actively encouraged and were part of a campaign of roadmindedness, giving public speeches, writing in the general press, and partaking in lobbying work.[25] While they strove to retain a focus on serving existing local or regional traffic, several other groups proclaimed extensive ideas for long-distance roads, especially after World War I.

Local, regional, and national boosters promoted roadmindedness on their terms in the interwar period. Chambers of commerce, tourism promoters, and automotive and construction interests introduced proposals for continent-spanning roads such as the Lincoln Highway, connecting New York and San Francisco.[26] Rather than publicly financed new construction, these ventures were built on privately organized interests and lobbies and lavish public relations. To some degree, they were merely existing highways decorated with new road markers. Others were but partially realized. The Lincoln Highway, for one, remained under construction for years, especially in the deserts of the American West. No matter: "Keeping the name before the public and a never-ending pressure toward the great objective" was the publicity goal of the highway's association.[27] The growing number of car owners were to be convinced that a highway connecting the two coasts would be worth their time and, eventually, tax dollars.

Automotive traffic, of course, was intensely local, but roadmindedness was to be a national issue. The road- and car-friendly President Wilson received "Highway Certificate #1" in exchange for a five-dollar donation to the Lincoln Highway Association. Other initiatives with similar qualities of imagination included a Dixie Highway, Yellowstone Trail, and Atlantic Highway.[28] Some 250 names sprang from the pens of imaginative promoters—too many, in the eyes of federal administrators, who saw confusion and began to implement a regimen of numbering rather than naming highways, which exists to this day.[29] Numerical or nominal, roadmindedness had been firmly planted.

The infrastructural enthusiasm in the United States far exceeded any similar sentiments in Germany before 1933. The central government in Berlin left roadbuilding up to individual states. Given the low levels of motorization, they paved some interurban roads and highways to cope with cars and trucks. Civil engineers and their organizations hailed roadbuilding as necessary and inevitable and a lobby for a national highway network emerged, as chapter two will show. But all of these efforts remained inconclusive.

Did roadmindedness extend to organized, self-professed nature enthusiasts? Conservationists, of course, used road and rail to explore the landscapes they cherished, but they were uncertain whether expanding access through easier transportation would be good or bad, for them and for nature. In 1901, John Muir, the founder of the Sierra Club and an eloquent apostle of the wilderness movement, was happy to observe that "all the Western mountains are still rich in wildness, and by means of good roads are being brought nearer civilization every year."[30] By 1914, an umbrella organization for the preservation of historic monuments and the conservation of nature had established a simple dichotomy between contemporaneous technology and the landscape.[31] Decrying the "mutilation and disfigurement of notable features of the natural landscape," the group opined that recent developments in the physical sciences and in engineering had led to a "commercial assault" on nature: "In older times the highways generally followed natural grades and curved around hills and other obstacles. Today, the engineer draws a straight line, and blasts his way along the shortest distance between two points. The highway and the railway defy Nature and go where they will."[32] This wholesale condemnation of "the engineer" would flatter the self-image of the educated middle classes as wardens of culture and nature; some of those engineers, as we will see, tried very hard not to live up to the stereotype. Road enthusiasts, in return, often praised roads as a new way to bring urbanites into nature, with one claiming in the mid-1920s that "folks who ten years ago were unfamiliar with grass except as it grew in parks can now distinguish instantly the difference between poison ivy and the trumpet vine."[33] Muir's surprising embrace of roads and the categorical rejection of road and rail as impositions by other nature lovers were two ends of a spectrum. As mostly middle-class urbanites, environmentalists needed infrastructures to reach their destinations, which were increasingly marked by the absence of such technologies. These tensions remained and contributed to the debates over and the design of scenic roads. Muir's statement points to a possible unity of roads and scenic exploration, which the National Park Service would later promote in much more pronounced ways. Roadmindedness provoked and contained contradictions.

Roadmindedness across the Atlantic

While such manifestations of roadmindedness appeared to be exclusively national, they were part of international networks and exchanges. State sponsors and planners eagerly celebrated roads and highways as vernacular, national, or even nationalistic achievements, yet professionals such as civil engineers, landscape architects, and urban planners were keenly aware of developments in other countries. They incorporated international design ideas and management techniques. In the process, they elevated roadmindedness to a mindset and practice whose core elements circulated freely across the Atlantic. Yet, it took on specific national forms whenever it was realized. This inherent tension between international exchange and national expression was not exclusive to roads and highways, of course. But the politics of expertise mattered in specific terms.[34] In the sphere of international knowledge circulation, roadmindedness was an interrelated process driven by experts who portrayed themselves as apolitical. Scientific and engineering journals, papers, visits, and international conferences were conduits for exchanging ideas, policies, and administrative procedures, and reinforced a collective identity of road planners as improving society by undergirding car-based mobility.[35]

One institution stands out: the Permanent International Association of Road Congresses (PIARC). It became one of the premier arenas for road experts from Europe and the United States to trade knowledge and practices through its publications and gatherings. PIARC meetings were elaborate, state-sponsored affairs with opportunities to compare plans, projects, publicity campaigns, and finished roads. Delegates, mostly civil engineers and government officials, represented their home countries, took notes on reports from other nations, and went on tours of roads in the host nation.[36] France, with its centralized planning and tradition of elite schools, had been prominent in producing and codifying knowledge on highway building since the early modern period. Therefore, it comes as no surprise that the first PIARC meeting was held in Paris in 1908. At this gathering, the delegates spent a lot of time worrying about the dust problem caused by urban automobiles in the countryside. Bluntly, the chief engineer for the city of New York rejected the "segregation of motor traffic" by "dustless" roads exclusively for cars. They would be too expensive to build, he argued, and make motoring unaffordable for most Americans, as the cost would have to be passed on to them through taxes.[37] For the American representative to the first transnational road congress, a sole-purpose road was a curiosity, not a solution to a traffic problem.[38] While his East Coast colleague castigated special roads, Samuel Lancaster of Oregon participated in the same conference, but he did not leave a trace in the conference proceedings other than his name. It is clear, though, that he sought and found inspiration in Switzerland, France, and Italy, where he studied coastal and lakeside highways during his European trip. Eight years later, the highly referential Columbia River Highway opened to traffic.

The 1913 PIARC meeting in London was marked by more calls for new roads. Sir George S. Gibb, chairman of Britain's Imperial Road Board and of the congress, noted the "startling suddenness" of the appearance of automobiles in his opening address. "The old roads can no longer satisfy the new needs," Gibb claimed, thus marking the transition of the Road Congress from a forum for collecting expertise to a group defined by engineering and political advocacy. It became a body that described an urgent problem and offered solutions to it in the form of more roads.[39] Cloaked in the language of expertise and disinterested advice, the proceedings of these congresses speak to a deeply political act of expanding roads and redefining them. According to one historian, PIARC became a lobby for roads.[40] Reports and presentations at the meetings encompassed a large variety of technical issues, while the overwhelming message became clear: the rise of the automobile called for massive investment in new roads. While other international organizations pursued similar messages, PIARC appears to have been particularly steadfast in its message of roadmindedness.[41]

By the time a PIARC congress came to Washington, DC, in 1930, a British delegate indicated that the publications of the US Bureau of Public Roads had familiarized non-Americans with stateside roads. Although it was the first visit to the United States for many delegates, they had already seen the country through the eyes of the Bureau.[42] Publications, in other words, had helped to establish the "highway fraternity" that Thomas H. MacDonald invoked when greeting the delegates.[43] A fraternity it was indeed, since female civil engineers, small in number as they were, were not represented among the delegates at all. During post-conference excursions, the sheer number of cars and the mileage of roads in the United States left a deep impression on European delegates. In professional terms, they also noted the comprehensive nature of road planning and the institutionalized expertise of road engineers. The Americans presented urban, suburban, and rural roads. Much to the astonishment of Europeans, traffic meant almost exclusively motorized traffic; all other kinds had vanished: "In the countryside, we did not meet any pedestrians, nor bicyclists, and horse-drawn vehicles were nonexistent. The roads are owned uniquely by automobiles," according to one astonished French reporter.[44] The federal government showed off construction sites of its signature scenic road of the time, the Mount Vernon Memorial Highway.[45]

Thinking about roads and debating their scope, design, and location had become a standard exercise for planners and politicians by the late 1920s, both in Europe and in the United States. Engineers and planners working at local, regional, national, and international scales preached and practiced roadmindedness. The roles of government and experts working for it changed to the point where Thomas MacDonald could claim with only some hyperbole: "Thus, the building of highways adequate in character and in extent becomes, next to the education of the child, the greatest public responsibility in all of our, otherwise highly developed nations of this Western Continent."[46] Pedagogy and pavement, in other words, were equally important.

The Culture of Roadmindedness

If roadmindedness was to succeed, it required cultural work, in addition to political and engineering efforts. At first glance, roads "scarcely admit of being treated in that easy, amusing, and instructive manner which less homely subjects might admit of," as one nineteenth-century writer put it.[47] Yet travelogues functioned as one mode of such cultural efforts. Before the interwar roadbuilding boom in the United States, and up until the 1930s in Europe, authors wrote in a jargon of discovery and danger about driving on existing roads, especially when traversing remote areas (in the case of North America) or the Alps (in the case of Europe). The resulting articles and books were entertaining and written for an audience accustomed to exoticizing tales, such as those of Arctic or African expeditions.

Sponsored by a magazine, the New York socialite and writer Emily Post—later to become famous as an authority on etiquette—set out for a road trip from New York to the Pacific in 1916 with her son. A book ensued. For her travels, she chose the most difficult mode, when a train would have whisked her across the continent in comfort. Post encountered a geography of difference. The "magnificent work" of car clubs or highway commissioners in the Midwest made travel easy. Given the absence of fixed roads and bridges in parts of the American Southwest, however, her son gave advice on fording streams.[48] The Lincoln Highway was only a chimera in some places. All in all, the picture of Post and her car parked on a Pacific beach made the point most vividly: traversing the United States coast to coast could be done, but only with difficulty.[49]

In fact, the American West was more exotic for the wealthy New Yorker than Europe. In her book, Post established her motorist bona fides by assuring her readers that she had "driven across Europe again and again" and claimed to have made it from the Baltic to the Adriatic in 1898 "in one of the few first motor-cars ever sold to a private individual. We knew European scenery, roads, stopping-places, by heart."[50] Post's assurances make her part

of a coterie of travelers and writers who practiced roadmindedness by driving some of the first automobiles on European roads.

One topographical feature between the Baltic and the Adriatic stood
out: the Alps. Their peaks and valleys, especially the Swiss ones, loomed
large over the cultural work of roadmindedness. Oregon's Lancaster was not
the only one to be enchanted and instructed by this range. In the history
of tourism and scenic infrastructures, Switzerland and its mountains figure
prominently. International cultural efforts at establishing roadmindedness
had an unmistakably Alpine quality to them, especially in the first two decades of the twentieth century.

Switzerland and Scenic Infrastructures

Roadmindedness in Switzerland was not simply a result of higher
mountains. Its scenic infrastructures were caught up in the country's early
and successful promotion of tourism. While this republic is certainly well
endowed with peaks of impressive heights, it did not become a vacation
destination without some other advantages—and some work. Switzerland's
relative proximity to major European urban centers, its eager embrace of
tourists, and its rapid buildup of scenic infrastructures contributed to its
becoming a favorite destination for travelers. After an intense period of railroad construction in Switzerland, neighboring countries built rail- and road-
based scenic conveyances as well.

Historians still debate how to chart these developments. Yet, it is clear
that Enlightenment "discoveries" contributed to opening the way for haphazard and then growing waves of tourists in Switzerland.[51] The British enthusiasm for the Alps is well known, especially in its mountaineering form.[52]
Gaining force in the mid-nineteenth century, climbers and hikers from Britain took to the Alps, supported by rapidly growing infrastructures. The
numbers of tourists were such that the *Saturday Review* decreed as early as
1867—prematurely, as it were—that the Alps were already "used up," as almost every peak had been climbed.[53] First ascents were no longer available,
but personal firsts were.

Jumpstarted by British mountaineers, who conceived of climbing as a
middle-class sport, and their Swiss hosts, Swiss tourism infrastructures soon
accommodated both the hardy individualist and the comfort-seeking traveler. The railway network grew quickly and extensively, reaching many of the
Swiss valleys. Tunneling ensured fast access; the planning and construction

of the Gotthard Tunnel in the late nineteenth century was celebrated as a technological feat, and as an icon of Swiss ingenuity.[54] While these trains followed the valleys and went through the mountains, cable cars, gondolas, and railroad branch lines enabled tourists to forgo the hike and leave the climbing to engines. Such mechanical ascents were both popular and controversial. As one historian argues, they both enabled and conditioned tourism, as some valleys and mountains became accessible by rail and others did not. Locals, after initial opposition, favored those mountaintop projects for the most part; conservationists who aimed to speak for the locals and for nature did not, in union with hikers and mountaineers; and banks and investors often decided whether or not plans became realized.[55]

The rise in the numbers of Alpine tourists and the rise in the height of peaks they could visit without major effort pleased most tourists and the tourism trade. However, two groups in particular objected to the ease of this scenic appropriation. Hikers and climbers could no longer look down upon day-trippers in a literal sense since they had to share coveted peaks with them, but their organizations did so figuratively. Mountaineering built character and human physique, they claimed, while the effortless ascent via modern transportation technology lacked authenticity and was a lesser form of scenic enjoyment.[56]

Together with hikers, conservationists, although often ambivalent about the effects of tourism on the Alpine landscape, were most predictably incensed by proposals for building cable cars and other means of easy access to mountaintops. In a 1908 publication, a Bavarian conservationist compared different ways of reaching the summits: "While it is an experience for the one who has made the achievement of reaching a proud summit through one's own strength, it is only a naked fact for the majority of those who let themselves be lifted up by steam or electricity." The latter mode was clearly trivial in comparison to relying on personal vigor and bodily strength, according to this observer.[57] Probably the harshest indictment of cable cars sprang from the pen of Ernst Rudorff, a composer and one of the founders of the German conservationist movement. Incensed by the construction of cable cars and mountain railways in Switzerland, he unsuccessfully sought a ban in Germany, where he claimed only a small minority, whom he called "traveling rabble" (Reisepöbel), would use them.[58]

Perhaps the epitome of such easy access to the peaks is the railway to the Jungfraujoch mountain in Switzerland, which still boasts Europe's highest railway station at an altitude of 11,332 feet (3,454 meters). Travelers en route spend most of their time in tunnels, only to be rewarded by high Alpine panoramas and creature comforts at the top. These tunnels were hugely expensive and very dangerous to build. This monument to technified mountains opened in 1912 and has remained a major tourist attraction. The Swiss and their guests developed a strong penchant for cog railways, cable cars, and luxurious, high Alpine hotels. The Swiss mountain landscape received extensive scenic infrastructures, thus making it a reference point for the promoters of roadmindedness.[59]

Even the railroad-loving Swiss did not overlook roads. For the most part, they left the lowlands to the rails and expanded or upgraded the existing road network in the mountains, where few Swiss lived but tourists dwelled. In the early 1860s, the Swiss national government decided to sponsor the construction of a cross-shaped set of four roads for commercial and military purposes.[60] One of them, the Axenstraße, became internationally famous because of its galleries and inspired the Columbia River Highway.

Such touristic success found both imitators and critics. In what set the tone for many publications to follow, the English climber and writer Leslie Stephen dubbed the Alps, and especially the Swiss Alps, the "playground of Europe" in 1871. It was, however, a playground with its own rules and marks of distinction. Stephen set up a stark contrast between mountaineers and "ordinary travelers," since the former were willing to exert physical efforts and take risks.[61] The rewards, then, were all the greater for those who climbed into "the farthest recesses" of the Alps: "And without seeing them, I maintain that no man has really seen the Alps."[62] For Stephen, where and how one visited the Alps was related to social status: "The bases of the mountains are immersed in a form of cockneyism—fortunately a shallow deluge—whilst their summits rise high into the bracing air, where everything is pure and poetical."[63] Mountaineering as a way of rising above the masses, both literally and figuratively, has been a quest for many of its practitioners since Stephen's times. Gender, class, nationality, and speed of travel mattered: Stephen mocked "ladies in costumes, heavy German professors, Americans doing the Alps at a gallop, Cook's tourists."[64] But the masses had a way of catching up with the Stephens of this world. The deluge turned out to be anything but shallow. More and more visitors came, and increasingly, they used scenic infrastructures rather than climbed.

Around the turn of the twentieth century, early motorists in the Alps answered Stephen's quest for authenticity with a cultural appropriation of their own. The savoir faire displayed by the urban middle and upper-middle classes when it came to food, lodging, and dress extended to motoring, and especially to motoring in difficult, mountainous terrain. Guidebooks and manuals recommended demanding roads (for the adventurous driver) or scenic ones (for the sensual driver of a slower bent).[65] If the difficult met the aesthetically rewarding, all the better. This roadmindedness has left a considerable legacy in guidebooks and testimonials.

Initially, these motorists traveled on existing Alpine roads, especially the ones traversing peaks. Alpine passes for connecting the lowlands north and south of the Alps were well established. By the early modern period, nine main passes had handled most of the transalpine traffic and commerce. Some were upgraded for more traffic in the early nineteenth century. With the growth of commerce and tourism, guidebooks effectively channelized trips on these roads and the views of these travelers, as the recommended avenues of sights were but few.[66]

Stories of such trips circulated widely. When automotive pioneers told tales about their trips, travel in the Alps was a favorite genre. One of the most widely read accounts in the German-speaking world was the 1903 travelogue by the writer Otto Julius Bierbaum of a "sensitive trip" from Germany to Southern Italy and back. Bierbaum claims that he (or more accurately, his chauffeur) was the first one to drive up the Gotthard Pass road in Switzerland in an automobile. It was one of the many highways built during the early nineteenth century for strategic and economic purposes, only to be eclipsed in significance by the railroads a few decades later.[67] Bierbaum reports that it took him all of nine hours to travel eighty-five miles (136 kilometers) of mountain roads, as he relied on a one-cylinder, eight-horsepower car. Rather than describing the mountain scenery in detail, Bierbaum's account emphasizes the slow pace of the trip. After being fined by a Swiss policeman for illegal driving and admonished to drive more slowly, the author added almost petulantly that he did not need the advice "as it would be a sin to hurry here."[68]

Such travel accounts promoted a cultural validation of roadmindedness. The bourgeois preference for unhurried driving reflected a distaste for the scheduled experiences of the train and the mixing of passengers from different classes. In the case of early mountain driving, the thrill of adventurism was added, as it was uncertain whether road and motor conditions (together

with the skills of the driver or chauffeur) would allow the trip to be completed. As late as 1958, the British writer Hugh Merrick classified the Stelvio road, another remnant from the early nineteenth century, as one that
should only be driven by "experienced drivers for its own sake."[69] Not least
because it was the "loftiest" in Europe, the Stelvio received more written
praise than most other pre-automotive mountain roads. Its hairpin turns
stood out.[70] Merrick was one of the most eloquent observers to praise the
sport of mountain driving:

> Wriggling and writhing, darting and dodging, now this way now that, by a con-
> tinuous bank of walled hairpins, now built out one above the other like the land-
> ings of a giant spiral staircase, now spaced out at the end of long straight sectors
> where it clings to exiguous ledges, the narrow white ribbon of this incredible
> road literally slashes its way over a height of 3,000 ft. of precipitous mountain-
> side in fifty looping, swirling bends which leave the beholder almost dizzy as he
> tries to sort out the interweaving pattern of its bewildering course.[71]

The pleasures and pitfalls of switchbacks have rarely been described more
evocatively. Especially before the 1920s, driving on these roads required skill
and constant attention to one's environs, challenged and reinforced the
sense of masculinity of these motorists, and rewarded them with a sense
of technological mastery and plenty of views. While roads in the lowlands
could be lovely, even picturesque, Alpine highways were demanding and
grandiose to the same degree.[72] As late as the 1950s, Merrick claimed, "The
main thrill is still there, and each crossing is still an undertaking and an
adventure."[73] He also compared Alpine driving with mountaineering, did
not find it wanting, and deemed it a "different but complementary form of
high-mountain travel."[74]

In these accounts, roadmindedness possessed a sportive quality. This
was especially true for roads such as the Stelvio whose raison d'être had been
commercial or military, and which predated the automobile.[75] Steep gradi-
ents and hairpin turns had suited carriages or animal-powered carts well
enough. Drivers of early cars could not always turn corners on the first try,
were wary of rolling down the hill when trying again, and could not always
rely on their brakes. Once the trip was completed without incident, writers
felt a sense of accomplishment and mastery.[76] Built in the early nineteenth
century by the Austrian Empire to connect its provinces of Tyrol and Lom-
bardy by carriage travel, the Stelvio, with its eighty hairpin curves on both

The Stelvio Pass. The switchbacks on the Stelvio mountain pass in Europe attracted early motorists looking for a challenge. They considered driving to be
a sporting activity. Charles L. Freeston, *The High-Roads of the Alps, a Motoring Guide
to One Hundred Mountain Passes*, 2nd ed. (New York: Charles Scribner's Sons, 1911), 231.

sides of the mountain, allowed motorists to reach an altitude of more than
nine thousand feet (2,757 meters). A 1911 guidebook claimed that "every
summer the road is traversed by cars in plenty." Its author also warned
drivers not to overheat their engines when driving, as doing so led to many
engine failures. But it was worth the effort and potential damage, since "on
no other road can such magnificent views be enjoyed."[77] Of course, tourists
could ascend mountain peaks with less skill and effort by taking trains and
gondolas, but motorists insisted that those methods were less sporting.

This group of motorists embraced uncertainty. In fact, the less predictable the trip was, the better. The first handful of years both before and after
1900 offered drivers plenty of chances to be first in ascending particular
peaks. The relatively weak engines, dearth of gas stations or repair shops,
and novelty of the experience turned the entire effort into the sport that
upper-middle-class pioneer motorists were after. Looking back in 1913, a
Frankfurt driver reminisced about a trip undertaken just thirteen years earlier. With a 4.5-horsepower engine under the hood, the trip to the top of the
Stelvio Pass had taken two days. "This was still a true sport," the writer re-

membered; his trepidation had made it so. He claimed to have been the first
motorist to ascend the Stelvio's switchbacks so eloquently described by Merrick. (It is safe to assume that these claims to fame outnumber the roads and
peaks of the Alps.)

Solitude in motoring was another feat: in 1900, while adding about 2,500
miles (4,000 kilometers) in northern Italy and Tyrol to his odometer, our
Frankfurt driver had encountered only one other car. By 1913, he claimed,
one would have to travel to the farthest reaches of the Balkans to escape
other motorists. The only relief after traversing a pass was realizing that one's
car had not been dented by another one. Making it across was now a given.[78]
Tales such as this one served to distance first-generation long-distance
motorists from later urban and suburban drivers.[79] They also put an experiential gap between early adopters and those of the second generation. The
former had to endure unpredictability; the latter could rely on larger cars
with stronger engines, better brakes, and more roadside services. What
united both was a preference, at least when writing, in framing the road
trip through the lens of landscape. In the process, they created their version
of roadmindedness.

In the interwar period, the German upper-middle-class motoring journal *Motor-Tourist* was especially outspoken in its advocacy of car-oriented
landscapes. In 1929, the journal sponsored a competition entitled "How do
We See the Landscape?" to enable contestants to explain in writing and by
taking pictures "how they are proficient to enjoy the beauties of cities and
countryside and to feel their cultural and artistic attractions."[80] The idea was
to describe what one saw while motoring. Seeing was the motorist's preoc-
cupation and calling, since they had to pay attention to traffic and surroundings. But there was more: "Images and images hurry past him. The world
becomes a scenery for him, he drives past [as if] in mid-air. In the morning,
[he is] still in his daily city routines, amongst gigantic mountain by noon,
and by evening at the ocean perhaps, at a lake, in the lowlands!"[81] The dynamic variety of the trip, with its smorgasbord of images seen through the
windshield, found its match in such breathless narration.

In such travelogues, references to two other twentieth-century technologies, aviation and cinema, were commonplace. Flying was out of reach even
for most of the wealthy readers of the journal, but the cinematic experience of
plot narration through imagery was affordable and had changed sensory patterns. Rather than simply replicating it when driving, these motorists aimed

to be their own directors and editors. Again and again, motorist-writers invoked the post-railroad privilege of automotive agency, of choosing to drive or to stop—whenever and however long they pleased.[82] Fittingly, one historian notes the "combination of cinematic perception while driving and of the proliferation of cinematic modes of seeing through movies," which affected mountain driving in particular.[83]

This individual mastery over the road trip rested on a different sense of time. An account of a trip over an Austrian mountain road stressed that motorists should not worry about being stuck in the mountains or not reaching their destination for the day. What mattered was the "vivid memory of this beautiful valley; he who does not take it along will regret it at home. But it takes time to hold on to this memory."[84] For these travelers, unhurried, deliberately slow travel was preferable to speedy jaunts as only the former would resonate on a deeper emotional level. While being able to go fast, these tourists and writers celebrated stretching out travel time. Such new temporalities existed side by side with and were a rejection of a culture of speed at racing events and on ever-faster trains.

Individual reports are filled with descriptions of the landscapes traversed, making the road trip a non-productive leisure activity for affluent motorists. In addition, the motoring club behind *Motor-Tourist* organized annual outings for which the slow-paced appreciation of local landscapes was the prime motivation.[85] While car races in the lowlands and on select slopes drew huge crowds appreciating the thrill and danger of speed, middle-class motorists traversing mountain peaks by themselves or in small groups celebrated unhurried travel.[86] These travelers collected road experiences and traded stories about especially difficult, especially remote, or especially scenic roads.

Tourism promoters in the Alps paid attention. Although they appreciated the well-heeled motorists and their disposable incomes, the idea of 1920s motorists on mid-nineteenth-century roads was not well suited to their vision of Alpine tourism. Local mayors and hotel owners—but also national parliaments—engaged in a veritable international competition over scenic roads. While the landscapes of tourism, in particular Alpine peaks, were presented as static and frozen in place, access to them varied greatly and changed rapidly. Early mountaineers relied on physical skill, good boots, and knowledgeable guides; later tourists utilized trains, cable cars, and increasingly roads, in an ever-growing panoply of technologically enhanced consumption choices available to anyone who could pay for them. This sprinkling of Alpine scenic infrastructures was highly controversial, with some Alpinists deriding the less-sporting class of tourists.

Some mountaineers, however, embraced automobiles and even built roads for them in their version of roadmindedness. When South Tyrol was still Austrian, the local section leader of the German and Austrian Alpine Association in Meran-Merano, a hotel owner named Theodor Christomannos, successfully persuaded the Vienna government and local sponsors to fund a "Dolomite Road" (Dolomitenstraße) leading from Bozen-Bolzano to Cortina d'Ampezzo over a distance of 70 miles (112 kilometers). It opened in 1909, when automobiles were still rare. Through tenacious lobbying, Christomannos found money to create a "showpiece" of both the Dolomite peaks and of roadbuilding.[87] While the road served primarily tourists, the road's promoters did not cease to point out the military relevance of such a connection. They turned out to be correct.[88] Once finished, the road was utilized during the fierce battles over the Alps in World War I.[89] After serving as a token of pride for Austrian tourism promoters, the highway became a beacon for Italian tourism as South Tyrol changed hands after the war.

This sense of nationally charged and technologically enhanced scenery was just as strong in the French project of the "Route des Alpes." Like all the other Alpine highway plans, its promoters made sure not to cross national boundaries, lest tourists be led astray. As early as 1909, the Touring Club of France, the country's organization of wealthy urban motorists, had commenced construction of a high Alpine connection from the French part of Lake Geneva all the way to the Mediterranean, over a distance of some 435 miles (700 kilometers). Existing roads and some newly constructed ones should form a branded mountain-to-ocean connection. The touring club's ally was a major railroad company. It saw the "Route des Alpes" as an opportunity to transport tourists in open-top buses that met them at nearby railroad stations. In addition, scenery and roadbuilding were to receive the blessing of the nation-state. Raymond Poincaré, the French president, had planned to lead the opening ceremony for a stretch of the road in August of 1914, but he had to tend to more pressing international matters. Instead of tourists, mountain infantry soldiers used the road and explored the peaks of Southeastern France during World War I. Construction recommenced when hostilities ended, and finally, one of Poincaré's successors, Albert Lebrun, opened the highest paved mountain pass in the Alps, the Col de l'Iseran in

Savoy, at 9,068 feet (2,764 meters) above sea level, as part of the Alpine route in 1937, after several years of construction.[90]

Not to be left behind, Austria presented several scenic roads to its citizens and to the world during the interwar period. The demise of the Austro-Hungarian Empire translated into a dramatically smaller territory with fewer Alpine peaks, given the loss of South Tyrol and Slovenia. To make the highest mountain of the country, the Großglockner (12,461 feet, or 3,798 meters), accessible with a high-altitude road was both a patriotic and technologically daring project, as one historian writes. Its main promoter, the civil engineer and mountaineer Franz Wallack, presented plans for this scenic infrastructure in 1924. They generated lots of attention and debate, fraught with symbolic meaning as the proposal was. The price tag for the project was extremely high, however. Only the authoritarian Dollfuß regime was willing to provide major resources for the road, which opened in 1935 with a maximum elevation of 8,215 feet (2,504 meters), as discussed in chapter three. It also sponsored the construction of a smaller, more accessible road in the vicinity of Vienna, the "Viennese High-Altitude Road" (Wiener Höhenstraße). National identity was to be caught up in technological symbolism with these projects, whose long planning periods were followed by many non-Austrian observers and inspired competitors and would-be competitors.[91]

All of these projects were decidedly national in origin, motivation, and meaning. Whether in Austria, Switzerland, France, or Italy, tourism managers were careful both to attract tourist traffic and to retain it within their country's respective borders. In addition to obvious commercial interests, issues of national identity were tied to mountain scenery and automotive access to it, especially in Austria and Switzerland. There, territory and tourism defined each other to a large degree. Roads could be built as forward-looking and economically sensible monuments, despite their high costs. The triumphalism of contemporaneous accounts (and of parts of the historical literature) masks the degree to which these projects were initially controversial. Occasional grumbling from observers regarding cost and purpose was buried by a wave of enthusiasm. In terms of their role for tourism, Alpine highways created unique destinations for motorists and, thus, reasons to visit. At the time of their planning and construction, the hoped-for visitors were middle-class tourists from European metropoles in their own automobiles.[92]

Accounts of these scenic infrastructures show how they relate to other forms of movement. In the 1930s, an English preservationist who preferred hiking to driving quipped that whiskey-sipping motorists were "all liver and no legs."[93] Such critiques of new modes of traveling and tourism are quite common in the history of mobility. Generally speaking, older forms of mobility—and especially hiking, since its technologies are as ancient as shoes—are seen as more authentic; more comfortable and faster tours appear to alienate the travelers from their surroundings and are imbued with a sense of loss.[94] In the case of cars, however, the most recent technology appeared to bring not a deeper loss of one's perceptive abilities, but rather a regaining of the sense of landscape.

The touristic roadmindedness in Austria and Switzerland, with its sporadic ventures and scenic goals, had a counterpart in the less exalted but more grounded kind of roadmindedness in the German states during the interwar period. The national government abstained from planning for larger road schemes, given the low numbers of cars on the roads. But engineers and local politicians, especially those from urbanized regions, began to think about reorganizing roads along national lines. The Berlin government eschewed financing new highways, but by early 1932, important trunk roads were classified as long-distance roads. Roads were ranked in a hierarchy of importance.[95] Civil engineers founded a technical clearinghouse and road lobby; one of its most visible members echoed MacDonald's claim of the economic importance of roads, but with a slight twist: a relatively poor country such as Germany could not afford the luxury of bad roads, these promoters claimed. Roads would stimulate the economy and their expenses would pay for themselves, they argued, as they tried to overcome the reluctance of politicians and administrators to invest in these infrastructures.[96] Very little came of these plans. As chapter two will show, such occasional and aspirational efforts received support only with the rise of the Nazi dictatorship, which threw economic caution to the wind.

Roadmindedness and Institutions: The National Park Service

European projects in the realm of scenic infrastructure, especially in the Alps, met observers, imitators, and competitors in the United States. Roadmindedness needed enthusiasts and cultural ambassadors; it also needed institutions and funding. The cooperative roadbuilding program by American states and the federal Bureau of Public Roads was based on a utilitarian version of roadmindedness. While Bureau engineers were deliberating how to build roads and where to put them, another institution of the federal government was much more sanguine: the National Park Service. During the interwar period, it put scenic roads near the top of its agenda, provided copious funds, and built such roads quickly and extensively. The appeal of new roads in scenic regions was not lost on American planners, tourism boosters, and cultural observers. In fact, European exemplars of such roads figured prominently in stateside discussions about the cultural politics of scenery and tourism. First, they were role models; by the interwar period, a movement for domestic tourism included an emphasis on homegrown roads; and by the 1930s, European roads were seen as imitators of American ones. For the American parkway movement, European Alpine highways were curiosities, with fewer democratic qualities than domestic roads, given the lower rates of motorization.

Despite their internationally circulating design features, such roads embodied ideas about landscape and nation. For the Park Service, scenic roads and parkways were to be an expression of Americanness. The idea that landscapes were building blocks for nationhood was, of course, no less foreign to Americans than to Europeans. Landscapes from the Hudson Valley to Yosemite, from the picturesque to the sublime, figured prominently in the nation's understanding of its role among nations.[97] Given the country's vast territory, agricultural riches, and westward expansion, artistic landscapes played a crucial role in the way urbanites understood less populated regions of the country. Historians have pointed to a direct link between the popularity of landscape paintings for middle-class urban households and the nineteenth-century movement to establish national parks in the American West.[98] They have also highlighted how railroads, as privately owned transportation companies, pursued commercial interests by linking population centers to the remote parks.[99] In some contemporaneous accounts, the majesty of Western landscapes would make up for the fact that the United States did not possess centuries-old cathedrals, castles, or city centers, which became markers of nationhood in Europe.[100]

Scenic roads embodied these ideas in a new form. An important specimen of this nationally charged practice of scenic driving was the Redwood Highway, among giant coastal redwoods in Northern California's Humboldt County. Voters approved a bond issue for the road in 1909; construction began soon after; one of its sections was built with prison labor; and the entire road opened in 1923. When planning other forested roads, the California Highway Commission typically received freshly logged right-of-way from the respective counties that had sold the lumber for profit. In the case of the Redwood Highway, however, the Commission asked for and received a swath of land with trees still standing.[101] It made for more impressive driving and a shaded sylvan experience. This (more expensive) practice of a incorporating a beauty strip found many imitators over the course of the twentieth century, to the point of cartoons mocking it by the late century.[102]

In the late 1910s, conservationists from the Save the Redwoods League were among the promoters of the road. The Redwood Highway would bring attention and visitors to their cause of protecting tree stands.[103] Stephen Mather, the first director of the National Park Service, had been among the founding members of the Redwoods League. Madison Grant, equally known as a conservationist and eugenicist, expected that rising numbers of tourists would appreciate redwood trees while driving; the Save the Redwoods League hitched its wagon to auto tourism.[104] The Redwood Highway was to highlight American nature and the ancient trees, the "purity" of which was especially important during a time of massive immigration and cultural anxiety among old elites. References to Europe were implicit, as the trees became stand-ins for an immutable American identity.

Being a tourist, in this regard, was more than an experience in relaxation. The "See America First" campaign of the early twentieth century sought to increase not only the number of domestic trips, but to raise a collective appreciation of nature and culture that was understood to be American.[105] "See Europe if you will, but see America first," exhorted an ex-governor of Utah and president of the Salt Lake City Commercial Club, in a particularly bold 1906 speech, in which he asserted: "Don't be hypnotized by weird tales of European travel. There is not an attraction in the Old World that cannot be duplicated and discounted by the phenomena of America." The United States was the more pleasant place to spend one's dollars, as well, as domestic tourists would not be "hounded to death by a horde of mendicants."[106] While clearly using the Old World as its reference point, Western tourism boosters claimed that the natural scenery of the United States surpassed that of Europe. Not only was an overseas vacation unnecessary because of superior (or at least equal) American landscape; exploring the West instead of traveling across the Atlantic was also seen as a patriotic act—imbibing

American nature instilled and reinforced American values which, by definition, European destinations could not.

Originally, the See America First movement sang the praise of the Western United States, scenically endowed with the Rocky Mountains and touristically embellished with the older national parks of Yellowstone and Yosemite, among several others. But spending one's vacation dollars—assuming one had any to spare—domestically, rather than abroad, became more than a regional catchphrase after World War I. Local tourism boosters all over America were all too eager to portray their destinations as patriotic, as well as accessible.[107] A veritable outdoor industry began to emerge in the interwar years. It sought to regularize and capture automotive tourists, who had been exploring the countryside in increasing numbers.[108] In addition to local and regional efforts to attract tourists, the federal government provided support in the form of a quasi-touristic federal agency, the National Park Service. Its vision of tourism in the interwar years featured cars and roads as the coming means of transportation and scenic enjoyment.

During this period, publicly visible federal roadbuilding in recreational areas more often than not bore the stamp of the Park Service. Founded in 1916, the agency oversaw national parks mostly in the American West, as well as various historical sites. Its dual (and often contradictory) mission was to preserve parks and monuments and to make them accessible. In the case of Independence Hall in Philadelphia, access meant giving tours and interpreting the history of the site; in the case of Western parks such as Yellowstone, which were hundreds, if not thousands, of miles from the population centers of the country, access meant building roads. Especially under the leadership of Stephen Mather, the Park Service championed the idea of getting tourists to its parks in automobiles. Although railroads had been crucial for the establishment of these parks in the nineteenth century, Mather pushed his young agency to transform the parks for the automotive age. Little, he argued, had been done "to enable the motorists to have the greater use of these playgrounds [national parks] which they demand and deserve."[109] This was more than a matter of logistics, of moving visitors to the sites—and sights—and back home. The goal was to radically alter the experience of national parks. Instead of arriving with others by train, staying in a lodge, and going on excursions, either on foot or on horseback, visitors would arrive in their own automobiles, as families or in other small groups, stay in National

Park Service–managed campsites or small cabins, and tour the parks, mostly while seated in their automobiles.

Mobile landscapes of a different sort emerged. The train trip and the guided tour were to be replaced by cars and road trips. Inside and outside of the parks, landscape was framed through the movement of automobiles. Improved highways would bring more visitors to the parks, thus resulting in their greater popularity and more calls for expanding the park system. Mather aimed to augment the scale and scope of his institution, believed in growth and visions, and neither wanted to nor could shake off the attitude of the growth-oriented industrialist that he had been before joining the federal government.[110] In the context of Mather's Park Service, growth and success meant more roads. "Making a business out of scenery" was the goal, not just of local and regional tourism boosters (as it had been in Oregon), but of a fledgling federal government agency with an activist leader. In a 1916 article, Robert Sterling Yard, Mather's publicity person, again referencing the Swiss example, exclaimed:

> We want our national parks developed. We want roads and trails like Switzerland's. We want hotels of all prices from lowest to highest. We want comfortable public camps in sufficient abundance to meet all demands. We want lodges and chalets at convenient intervals commanding the scenic possibilities of all our parks. We want the best and cheapest accommodations for pedestrians and motorists. We want sufficient and convenient transportation at reasonable rates. We want adequate facilities and supplies for camping out at lowest prices. We want good fishing. We want our wild animal life conserved and developed. We want special facilities for nature study.[111]

Although Yard would later regret such sentiments and join the wilderness movement, Mather pursued landscape embellishments via roads, roadside parks, and observation points—to name but three accoutrements—with great vigor. As one historian puts it aptly, "Through the promotion of tourism in the National Parks, scenery itself became a kind of commodity."[112] In the words of another historian, a "windshield wilderness" emerged.[113] The design, production, branding, and promotion of this commodity was to be firmly in the hands of the Park Service, a touristic agency on a mission.

One of the most visible examples was Mather's support of a circular "park-to-park highway" that would connect Glacier, Yellowstone, Grand Canyon, Yosemite, and Mt. Rainier National Parks in a grand loop. While local

Park-to-park tour on the auto log of Sequoia National Park. With the support and participation of Stephen Mather, the first director of the National Park Service, a group of highway boosters conceived of a park-to-park highway connecting several national parks in the American West. On a publicity tour to seek attention for their cause, the group stops at Sequoia National Park in California and poses on its auto log, a fallen tree converted to a ramp. A. G. Lucier Collection, John T. Hinckley Library, Northwest College, Powell, Wyoming

boosters, including chambers of commerce and tourism managers, advocated and advertised this route, the states along the route were hesitant to pay for new roads for this purpose alone. The National Park Service had jurisdiction only over its own parks. However, this did not stop Mather from loudly and prominently supporting the idea of the park-to-park highway.[114] In a letter, he claimed to have come up with the idea himself in 1915, but left the public promotion to Western boosters.[115] With such a road in place, motorists could visit more national parks in one trip "without hardship," thus boosting visitation numbers.[116] In 1922, he supported plans for a national highway system spanning the entire United States, based on his observation that "travel is based on the enjoyment of scenery."[117]

This view contrasted with the more utilitarian motivations of the engineers of the Bureau of Public Roads, who favored roadbuilding to alleviate existing congestion and further commerce.[118] This meant roadbuilding in and between urban areas, which for Mather and most of his contemporaries were not scenic by definition. Building roads between cities might or might not lead truck drivers through scenic landscapes, but this was not the main point of consideration for the Bureau engineers.

The Park Service's expansionist road agenda, however, resulted in an increasing demand for professional experts. Figures such as Samuel Lancaster, a civil engineer moonlighting as an informal landscape architect, were a rare breed by the 1930s. At any rate, the academic training of either profession often discouraged rather than encouraged such interdisciplinary work. Landscape architects sought to distinguish themselves from architects, on the one hand, and mere garden design, on the other hand, in professional terms by stressing the artistic and comprehensive planning quality of their work, while civil engineers of the kind employed by the Bureau sought to distinguish their work by their use of quantitative research methods.[119] This is not to say that only landscape architects would know how to fit a road into the landscape, or that only civil engineers would know the appropriate curve radius or gradient of a road—far from it. Rather, by the 1920s, expertise over these matters was, to a large degree, a question of drawing professional boundaries and, subsequently, finding properly defined common areas.

Given this background, historians have remarked upon a 1926 interbureau agreement between the National Park Service and the Bureau of Public Roads as a turning point for scenic roads in the United States.[120] It was more than simply an accord for pooling resources from both parties for the purposes of road planning and construction. Codifying existing cooperation at Glacier National Park and elsewhere, this contract gave the Park Service control over the questions of where, when, and how park roads would be built. The Bureau was responsible for surveys and for providing building specifications in contracts for private companies, which it also supervised. Historian Ethan Carr argues that this agreement "structured decades of cooperation between the two federal bureaus."[121] It also cemented the predominance of the landscape architects and the Park Service and relegated the Bureau to a secondary role. Professionally and organizationally speaking, landscape design and landscaping roads in national parks became the domain of landscape architects, with aid from civil engineers for the latter. Although Bureau engineers might have disliked parkways in general because of their prohibition of common-carrier traffic, working with the Park Service on park roads and parkways gave them an opportunity to plan and build roads when a national interstate highway system had only the slimmest of chances of being funded by Congress.

The National Park Service promoted and built roads with scenic features extensively in its parks. It was also instrumental in developing the idea of parkways on a national level. These distinct roads feature prominently in the history of scenic driving. Limited to automobiles, their function was not simply to transport drivers and passengers, but to immerse them in the landscapes surrounding the road. More than any other federal agency, the Park Service embraced parkways. These types of roads gained prominence in the interwar period, with their divided traffic lanes, exit and entry ramps, and avoidance of at-level crossings. Their pedigree pointed to urban design and civic planning. The landscape architect Frederick Law Olmsted first coined the term "parkway" in 1868, in conjunction with his plans for Prospect Park in Brooklyn, New York. Primarily built for carriages, it had as few intersections as possible. It was designed as the unity of roadbed and adjacent trees and shrubs, as a "narrow, elongated park." Neither commercial traffic nor trolleys were allowed.

The meaning of parks underwent important changes in the nineteenth century, as the historian David Schuyler has argued. Instead of urban "associational and educational" spaces, parks were increasingly conceived as a "naturalistic landscape."[122] Parkways ensured the proper aesthetic movement through these naturalistic spaces, removed as they were from commercial activity and productive areas. These principles were maintained as the parkway—a way through the park or from park to park—became increasingly used for automobiles. The historian Clay McShane argues that the prohibition of common-carrier traffic on the parkways assured class segregation as well as the appropriate natural feel; social and environmental decisions were intertwined in the history of these roads.[123] Increasingly, some of the design features of urban parkways were utilized for extra-urban parkways. Large rights-of-way enabled planners to physically separate and to visually screen the roadway from surrounding areas. The road itself was adapted to landform through a curvilinear alignment that preserved scenic

Automobile advertisement in the *Ladies Home Journal* (1917). A female emblem of nature beckons motorists to leave the city behind and to drive in the countryside. While the choice of a woman to represent nature does not surprise, picturing a female at the wheel was indeed unusual for that time. Whether male or female, urban drivers used automobiles for jaunts to rural areas when cars were not typically used for commuting. The Henry Ford

features, such as streams and hills. Also, parkways introduced the idea of limited points of access, separate alignment for lanes running in opposite directions, and amenities such as roadside parks. Billboards and unchecked roadside development were the archenemy of parkway planners: hot dog stands with garish advertising became proverbial in the planning literature as examples of unsightly and unwanted intrusions into the landscape.[124] Instead, the idea of the parkway was to gain as much control over the road and the roadside as possible—an idea that led to many conflicts.

Major metropolitan centers such as Detroit, Minneapolis, New York, and Chicago extended their network of urban and suburban roads throughout the 1920s, whether they were urban boulevards, parkways under the jurisdiction of local park authorities, or federally funded state highways. To motorists, of course, these classifications made little difference. Planners expected and encouraged movement in and out of the urban centers, as well as traffic between suburban areas, thus creating a spiderweb of roads. Driving for pleasure became both possible and popular; in 1930, a Chicago area planner stated that the "only days of real congestion" on roads in the region were Saturdays, Sundays, and holidays. "Much of the traffic has no definite objective, and is contributed largely by persons out for a pleasure drive in a general direction from one to five or six hours."[125] Such a statement is all the more remarkable given the economic context: apparently, enough Chicagoans could afford to take their automobiles for recreational rides to cause traffic jams during the onset of what became known as the Great Depression. Pleasure driving had become an amenity in the more affluent urban and suburban parts of the United States.

As the extent of roads grew and their cultural meanings changed, roadmindedness became firmly established. Many of the road-centered narratives of the interwar years make roads into subjects of monumental importance that need to be rescued from ignorance: "The Road [sic] is one the great fundamental institutions of mankind. We forget this because we take it for granted," exhorted the French-English writer Hilaire Belloc in a treatise published in 1923 and sponsored by the British Reinforced Concrete Engineering Company. Belloc stated that, after many changes to and emanating from the highway, another turning point was now upon his countrymen. Melding a historical survey focused on England with a look toward the future, Belloc concludes with a blatant example of technological determinism: the internal combustion engine "will compel us to new roads," and they would have to be arterial and reserved for automobiles. He appears to be certain about the powers of this technology.[126] Belloc's output as a writer made the road treatise disappear under a torrent of other, more controversial publications; but it stands as a witness to the ways in which interested parties could create a new awareness for "the road" as an issue of cultural and political significance in the interwar period. It was sufficiently novel and noteworthy for a writer of Belloc's stature.[127]

Through the work of its literary, political, and administrative champions, roadmindedness had established itself as a marker of modernity by the 1920s, in both Germany and the United States. It was aspirational in the former and resulted in extensive planning and construction in the latter. Based on the international circulation of knowledge, the relationship between scenery and infrastructures was well examined, well documented, and increasingly well funded, at least in the United States and a number of Alpine locations. As was the case with tourism in general, Switzerland proved to be the forerunner and reference point for scenic infrastructures. Without fail, promoters of similar efforts elsewhere invoked Switzerland and its many edifices for more than a century. The small Alpine country became synonymous with technologically enhanced access to mountains and scenery in general.

While roadmindedness grew in significance, different groups and institutions pursued varying agendas and overlapping but distinct ideas. As a lobbying effort by interested parties such as civil engineers and landscape architects, it was an effort in professional politics and in establishing new and growing areas for employment. Roads, and especially scenic infrastructures, acquired new meanings as well. Rather than an ancient institution serving humans and animals on the move by themselves or in various vehicles, roads came to embody a twentieth-century version of industrial modernity in the realm of transportation: automobiles and new highways. They promised a cleaner and less hurried version of moving about, one that allowed for immersion *in* nature rather than speeding *by* it. In the eyes of their promoters, scenic roads could mend what railroads had ruptured.

Scenic infrastructures were not new in the interwar period, but the focus on scenic roads in the United States and Germany was. The touristic appropriation of scenery featured prominently, which helps to explain the outsize importance of Switzerland. But the aims of the interwar roadminded movement were much bigger. Its goal was not just to present beauty spots in isolated locations, but to remake the relationships between humans and the environment by implanting roadmindedness firmly, and by creating even more extensive scenic road infrastructures.

2

Roads to Power

During World War I, two landscape architects, an American and a German, served the armies of their respective countries in France: Gilmore D. Clarke (1892–1982) and Alwin Seifert (1890–1972). They never met on the battlefield. Instead of fighting with weapons, both helped to design military infrastructures as part of their service. After the war, they went on to design roads in their home countries and achieved considerable public recognition for their efforts to blend highways into the surrounding scenery. The degree to which they led parallel lives is remarkable. In the United States and Germany, Clarke and Seifert are most readily associated with the twentieth-century idea of marrying roads and landscapes, with landscape architects officiating at the scene. In political terms, their careers diverged dramatically: Clarke is best known for his design of regional parkways in the Northeast of the United States during the interwar period, and for participating in both the rise and the fall of Robert Moses's public works projects in New York; the high point of Seifert's career was his involvement with the Nazi dictatorship's projects of the autobahn and the German Alpine Road, where he rose to quick but limited influence for the entire Reich. Ultimately, Clarke's designs were born and negotiated under the auspices of a democracy, while Seifert had no qualms about tying his career to a dictatorship.

Clarke and Seifert were the most prominent figures associated with scenic roads in their respective countries; the scenic infrastructures that they championed and planned became some of the most visible and widely known exemplars of roadmindedness. These roads were exclamation points on the landscapes they traversed. According to their designers and acolytes, drivers and passengers on such roads would be able to immerse themselves in scenery, thus gaining a new appreciation of their surroundings. As picture-perfect as they appear, however, these scenic infrastructures often erased layers of human work on the land, especially that performed by persons of lower status. In some cases, remaking landscapes meant displacing locals and obliterating their dwellings.

Roadmindedness and its institutional carriers spanned the Atlantic and connected countries such as the United States and Germany. It is all the more remarkable, therefore, to realize that scenic roads in either country appeared in the garb of the vernacular. In terms of design, both Clarke and Seifert disdained the abstract modernism of the Bauhaus school and sought to counter it with regionalist, landscape-oriented patterns for road and roadside. For them, the degree to which their plans reflected and furthered ostensibly innate American or German values of landscape mattered greatly. Their public pronouncements speak of individual or national achievements, not international developments: their points of reference were mostly their home countries, where the purpose, design, extent, and use of landscaped roads were highly contested, to the point that road development became one with politics. In this sense, Clarke and Seifert were political actors. Both used the appeal of automobility for large-scale efforts to transform and re-design the landscapes of the United States and Germany. Together with millions of their compatriots, they embraced modernity in the form of the automobile but also sought to use its dynamism for reestablishing natural connections: in their eyes, ligaments of landscape—of forests, rivers, and open land—needed restitching in the form of roads, as did emotional tethers between twentieth-century dwellers and their environments.

At least initially, Clarke and Seifert trusted in and tried to establish the curative powers of the car-road complex. Early twenty-first-century observers might find such a stance to be odd at best, but these two architects and spokesmen for the landscaped road found themselves and their work in the mainstream of cultural attention, political power, and economic resource allocation. In doing so, they relied upon decades of political and cultural work by planners, automotive enthusiasts, tourism boosters, and writers who made the union of landscape and roads possible. Clarke and Seifert, in other words, both promoted and benefited from roadmindedness.

While historians of planning and architecture have examined the roads associated with these two individuals, it is worthwhile to assess their plans and the contexts in which they operated in a more comparative mode. Given the extent of parkways in the United States during the interwar period and the attention paid to them internationally, America figured prominently. Designers sought to emancipate themselves from European examples and aimed to create suburban drives without adventures or major driving risks, but with copious scenic intake. This type of road had no counterpart in Germany during this period, either conceptually or on the ground. German planners, however, were intrigued by the aspiration to counter the unplanned railroad journey with a professionally designed landscape of scenic surplus. While Clarke and Seifert stand out as specimens, the comparative envirotechnical history of roadways transcends their individual biographies. As the following pages will show, landscape architects were keen to assert themselves as design professionals during the first half of the twentieth century, with roads and especially parkways one of their most visible work sites.

Urban and suburban parkways in the vicinity of New York attracted considerable notice, both in the United States and in Europe, as they sought to provide a landscaped automotive immersion into scenery. Planning and building these landscapes, however, also involved displacing locals. Such social and environmental cleansing efforts were sometimes motivated by eugenicist thinking. Outside of the Northeast, other cities and regions spent resources on parkways and roadside improvement. German planners mostly observed rather than built such roads during this period, as the relative paucity of automobiles made new road construction economically questionable. Brushing aside such constraints, the Nazi dictatorship sponsored the planning and construction of the extensive autobahn network. While borrowing parkway rhetoric, these roads represented a jumble of propaganda, haphazard planning, and haste. In both Germany and the United States, design, scenery, and politics were deeply intermingled.

Landscape Architecture and Roadmindedness

Tellingly, both Clarke and Seifert performed the work of civil engineers during World War I, without having received formal training in this discipline. In the spring of 1918, Clarke designed bridges in the battlefields of France as a member of the Army Corps of Engineers, where he earned the lifelong nickname of "Major." His units supported the Allied war effort by building heavy steel bridges over the River Somme, its tributaries, and the Somme Canal as part of the successful efforts to block a German offensive in March of that year.[1] Some 185 miles to the east of Clarke, Seifert had planned and supervised the building of a military light railroad in Lorraine several months before Clarke's efforts bore fruit. Seifert adapted the tracks to landforms; the goal, however, was not aesthetic gain, but camouflage.[2]

For Seifert, blending nature and technology became a peacetime profession after the war. Born and raised in Munich, he had studied architecture at the technical university of his hometown. He took over his father's small construction company upon returning from France. After it went bankrupt he found unsteady employment as a freelance architect and by teaching architecture classes at his alma mater. An avid gardener with an abiding interest in organic farming, he immersed himself and became part of the nascent field of landscape architecture. Several private backyards and gardens bore Seifert's imprint. But it was not until 1933 that he began to apply his ideas to large infrastructural projects.[3]

For Gilmore Clarke, landscape architecture had already been a profession before his military service; public works, in particular parkways, became his métier when he returned. His parents owned a nursery in New York City and sent him to a private school that prepared him for his studies at Cornell, first in architecture, and then in landscape architecture.[4] Clarke was part of the first generation of American-trained landscape architects. Rather than studying in Europe and visiting canonically designed landscapes, as most of the few garden and landscape architects in the United States had done before him, his training was entirely American.

Landscape architects on both sides of the Atlantic were keen to demonstrate the relevance of their new profession by participating in the technological transformations of landscapes in the twentieth century. Rather than arguing over whether new roads, power lines, or hydroelectric plants were necessary or desirable, their basic instinct was to search for aesthetically acceptable solutions to design challenges. This approach set them apart from conservationists and preservationists, whose critique of modernity was often more fundamental. To be sure, the lines between preservation, conservation, and landscape architecture were sometimes blurry: Frederick Law Olmsted's well-known 1865 report recommending the creation of a park in Yosemite is but one example of a landscape architect acting in the capacity of a preservationist. In Europe, architects and landscape architects sometimes also argued for nature preserves. Seifert, for example, was a prominent member of conservationist organizations in his native Bavaria. Preservationist goals coalesced with professional politics: as parks in America and

Western Europe were built on the idea of access by visitors, they needed transportation corridors and amenities. Designing master plans, roads, hotels, and campgrounds became the province of architects and landscape architects; the infrastructure of conservation and tourism depended on professional design experts, as the discussion of the National Park Service in chapter one has shown. Growing numbers of landscape architects found employment with the Park Service and dominated the institution by the 1950s.[5]

The Bronx River Parkway

Among those experts, Clarke was foremost in the United States, at least when it came to roads. After his graduation from Cornell in 1913 and other jobs, he began to work for the Bronx Parkway Commission, which had been set up in 1906 but was languishing for lack of funds.[6] The Commission's twin goals were building a landscaped road and, in today's parlance, ecologically restoring this river valley in the Bronx and Westchester County. The primary method to achieve these goals was building a road exclusively for automobiles in a landscape cleansed of weeds and undesirable residents. The road had entry and exit ramps and no intersections at grade level. In the eyes of middle-class professionals such as Clarke, the Bronx River Valley had deteriorated socially and ecologically. Recent immigrants, many of them Italian Americans, lived in unsightly shacks, blighting the surrounding landscape, according to these views. For the promoters of the project, building a transportation corridor in a rejuvenated environment was both a return to a more idyllic landscape past and an embrace of modernity in the form of the car-road complex. Clarke assisted the German-born Herman Merkel, a trained forester employed by the New York Zoo, who consulted on landscape matters for the road. After the war, planning and construction of the Bronx River Parkway accelerated and the road was opened in 1925. Although a mere fifteen miles (twenty-four kilometers) long, the parkway garnered publicity both in the United States and abroad.

Photographs of the Bronx River Parkway were one of the most effective ways to promote the road, both domestically and internationally. The commissioners and planners for the parkway sought to obliterate what they saw as an overly commercialized, polluted river valley inhabited by poor immigrants, and replace it with a picture of scenery and purity. In their eyes, the billboards put up for railroad passengers and the unkempt valley floor

One of the sites for the Bronx River Parkway. With "before" images such as this one and the "after" photo on the facing page, the Bronx River Parkway Commission portrayed its planning and building efforts as salutary and scenic. The human costs of displacement were invisible in these photographs, which contributed to the outsize attention that this short road received in the United States and Europe. Library of Congress, Prints and Photographs Collection LC-J717-X98-52

with garden plots needed expurgatorial professionals, not the gardening efforts by immigrants. The result was a remade river, newly planted trees and shrubs, and a placid park scenery. The railroad embankment can be seen on the extreme right of both pictures. Historian Timothy Davis calls these practices "deliberate and often deceptive." Residents were never shown, only their dwellings, which did not live up to middle-class standards.[7]

Beyond the Bronx River Parkway

With visibility from the Bronx River road, Clarke went on to design parkways on his own. Robert Moses's patronage catapulted Clarke into an elevated position of nationwide public advocacy for parkways. He acted as a culturally versatile ambassador who could imbue these highways with cultural prominence in a rapidly changing country. He gained considerable

The same site after the construction of the Bronx River Parkway. Library of Congress, Prints and Photographs Collection LC-J717-X98-53

prominence by codifying the design language for parkways, and by asserting their importance among architects, writers, and politicians. Clarke was a self-assured expert championing a culturally charged infrastructure for a new technology. The growing presence of automobiles in the 1920s made questions of roads and road design pressing policy and design issues, and Clarke made sure that his voice was heard and his plans implemented.

As diminutive as it was, the Bronx River Parkway was celebrated as the first automotive parkway in professional and popular publications inside and outside the United States. Dozens of publications praised its novelty and virtues. In terms of design, this was one of the first parkways outside a park and outside a city. The designers incorporated curvilinear alignment and, first and foremost, an emphasis on the experience of the ride. Grades were separated through bridges, underpasses, and on- and off-ramps, thus ensuring an uninterrupted drive. In addition, the planners controlled vistas for the drivers and land use for local residents: the parkway's right-of-way was an astonishing 600 feet (183 meters) wide, on average. Billboards and houses were erased from the landscape; a more open, naturalistic design predominated.[8]

Aerial view of the Bronx River Parkway. Seen from above, the Bronx River Parkway's undulating design contrasts with the straight railroad. The wide right-of-way of the road includes the remade Bronx River Valley, whose inhabitants were relocated. Courtesy of the Westchester County Archives

The Westchester County parkway planners were not at all shy about tooting their own horns. In a report to the 1930 International Road Congress in Washington, DC, Jay Downer, a Princeton-trained civil engineer and chief engineer of the Westchester County Park Commission, highlighted grade separation through bridges and underpasses as one of the achievements of what he termed the "Westchester county type of traffic parkway." At an average cost of $100,000, such bridges and underpasses were expensive items. In contrast to regular highways, the Bronx River Parkway prevented private landowners from accessing it other than by using entry ramps through legal and design means. "Reservational control" was the goal.[9] Downer's account contributed to the outsize reputation that this road acquired in the United States and Europe. Delegates at the Congress partook of the publication frenzy surrounding the Bronx road. Professional journals and general-audience newspapers praised the novelty of its design.

Constructing a Landscape of Control

Scenic driving and ecological restoration coalesced in this project.[10] The third interwoven element was social control: some residents of the Bronx valley were forced to move out of what the planners called the "reservation." Their houses were auctioned off and demolished. The dwellings of these poor, working-class Italian immigrants did not align with the envisioned scenic qualities of the drive. Once the parkway was completed, no buildings were allowed within three hundred feet of the road.[11]

This type of roadmindedness rested on environmental and social control. The drive to the northern New York suburbs was naturalized to the extreme on this parkway, with plants and trees shielding suburban development from passengers and drivers. The designers extolled the ways in which both the natural and the social environment of the Bronx River Valley were restored through the planning and construction process: instead of weedy and unkempt waterscapes and occasional garden plots, a pleasing, unproductive, and complete landscape could be seen through the windshield. From a design perspective, the pictures presented aligned with the picturesque, English style of landscapes, in which undulating paths met the pleasant view of meadows and valleys.[12] In an oral history from the 1960s, Clarke noted that some design parameters, especially for bridges, were adaptations of European models.[13]

The constructed landscape—the total volume of earth to be moved was two million cubic yards—was at once picturesque and decidedly modern. A cleaner river and valley provided the backdrop for drivers and passengers. Modernity also meant the control of scenery and social setting. One of the three commissioners for the Bronx River Parkway was Madison Grant, whose equally passionate advocacy for conservation and eugenics has made historians ponder a causal link between the two. Grant's book *The Passing of the Great Race* lamented the decline of "Nordic" European stock and became a staple of voices seeking to restrain Southern and Eastern European immigration to the United States.

For conservationists such as Grant, advocacy for protecting landscapes went hand in hand with a desire to control population and reproduction, undergirded by clearly racist views. He advocated for preserving redwood trees in California, for immigration control nationwide, and for the Bronx River Parkway. In all these endeavors, he "pursued a vision of sanitized, managed landscape as the moral environment needed to combat the degradation of American culture," according to one author.[14] For Grant, redwoods in the West needed protection from the onslaught of modernity, as they represented a pure and primeval America. In the East, reworking a valley floor by moving undesirable residents, cleaning up a waterway, and building a parklike road purified the land with the help of modernity's accoutrements: professional planners, pavement, cars on the move, and controlled vegetation. A return to an imagined prelapsarian natural state was the goal.[15]

The apparent appeal of early parkways aligned with the exclusionary goals of some of their proponents in this interwar version of roadmindedness. The pleasure and beauty associated with a drive on the Bronx River Parkway depended on excluding untidy people and landscapes. Ordering, classifying, and remaking was at the heart of these ideas. As the historian Alexandra Minna Stern observes, the "apparition of eugenics sits restlessly at the heart of American environmentalism."[16] In the realm of American suburban parkways, displacing people and remaking landscapes merged in the name of beautified landscapes.

Three lasting legacies arose from the Bronx River Parkway: the emergence of a vocabulary of beauty and accessible nature, the professional coalition of civil engineers and landscape architects, and the realization that property values alongside the parkways increased.[17] Therefore, local governments could justify the expenditures for these types of roads—higher property taxes would result. Between 1923 and 1933, New York's Westchester

County spent over $80 million to complete a system of parks and parkways. Even more (in)famous is the work of Robert Moses (1888–1981), as the chairman of the Long Island State Park Commission and, from 1934, as Commissioner of Parks for New York. Gilmore Clarke was his chief landscape architect. Together with the Cornell-trained landscape architect Michael Rapuano, Clarke designed many of the Depression-era parkways associated with Moses. In 1937, Clarke and Rapuano established a consulting company for engineering and landscape architecture, which not only offered designs for parkways and expressways, but also community master plans, programs for urban renewal, and planning for college and university campuses.[18]

Robert Moses in the Landscape of Transportation

Moses looms large over the history of transportation and mobility in the United States. Needless to say, he was roadmindedness incarnate. His outsize persona, political acumen, longevity, and advocacy of car-friendly cities and environments have since made him anathema to community-oriented planners and critics of automobility who see in him the epitome of high-handed, top-down decision-making. His most prominent biographer argues as much.[19] One of the last vestiges of Progressivism, Moses changed the landscape not only of the city and state of New York, but also of planning and roadbuilding nationwide. His downfall in the 1960s was as spectacular as his rise during the first half of the century. Without ever holding elected office, Moses amassed and defended power for several decades. His loss of political loyalty with a new generation of politicians and the emergence of community activists such as Jane Jacobs contributed to his undoing. Playgrounds, beaches, bridges (including the massive Triborough Bridge project), and roads bore his imprint. Scenic infrastructures helped to establish his name nationally. Apart from his personality, the ways in which Moses marshaled expert knowledge and molded infrastructures embody a confident, mid-twentieth-century professionalism that brooked almost no opposition. More recently, some historians have painted a different picture of Moses by accentuating his "effectiveness within a system of constraints."[20] No stranger to controversy, especially in his later career, Moses remains a subject of disagreements.[21] One anecdote in particular speaks to these qualities. A parkway on Long Island leading to Jones Beach, itself enhanced under Moses, features bridges with clearances so low as to block buses, which would have carried predominantly African American visitors without automobiles—

segregation in concrete terms. Or so the story goes; Moses's biography and an oft-cited academic paper have spread it widely. Some scholars disagree, noting oversimplification and even including a reproduction of a bus schedule in a learned paper. Regardless of its underreported complexities, the account and its durability are instructive. Given how polarizing Moses was, how he generated equal amounts of veneration and abhorrence, it is perhaps no wonder that the popular notion has survived for so long.[22] It does, however, divert attention from the more important observation that the Moses parkways were built when most households in New York City did not own an automobile, thus creating exclusion on a much broader level than a set of bridges. Moses and his planners had no qualms about razing neighborhoods and displacing residents both for parkways and expressways in bouts of infrastructural racism.

Seen in this light, it is remarkable how the imperious and abrasive Moses helped to put road systems into place that made highbrow critics swoon over their beauty. Memorably, Moses himself characterized his approach to urban roadbuilding after 1945 by saying, "When you operate in an overbuilt metropolis, you have to hack your way with a meat ax."[23] His interwar plans for suburban roads might perhaps evoke the image of him wielding a scalpel instead. His sharpest postwar critic, the architectural writer and public intellectual Lewis Mumford, gushed over the marine parks and the "great landscaped highways" leading to Long Island beaches in the 1930s. They left him "in a state of ecstatic admiration." Mumford was impressed by the design and the careful selection of plants and shrubs.[24] The Taconic State Parkway, a northern extension of the Bronx River Parkway in the Hudson Valley, lived up to the level of a "consummate piece of art" in Mumford's eyes.[25] Unperturbed by the regionalist architectural styles on parkways, the modernist architectural critic Sigfried Giedion professed being exhilarated by simultaneously "being connected with the soil yet . . . hovering just above it" when driving on such roads. Parkways embodied a specifically modern appreciation of space and time for this writer: "The space-time feeling of our period can seldom be felt so keenly as when driving, the wheel under one's hand, up and down hills, beneath overpasses, up ramps, and over giant bridges."[26] The novelty of an uninterrupted ride on highways designed for cars and the scale of the roadscape equaled modernity itself for Giedion. Modernist architects, unsurprisingly, embraced cars. In one of the most remarkable examples, Frank Lloyd Wright presented a plan for a mountaintop facility for a privately owned mountain outside of Washington, DC in the mid-1920s. The building was designed for automobiles, whose drivers would ascend on a broad ramp, enjoy views from the top, and then descend. Hotel rooms, a planetarium, restaurant, and other facilities would have completed this building. It was, however, never built.[27]

Proposals and voices such as these extolled the attention to detail and the designers' efforts. Drivers who did not contemplate how the road was made—probably the vast majority of them—were more likely to experience a recreational ride in a natural setting. Whether or not the landscape was contrived mattered less to them than the possibility of traversing terrain pleasantly in a non-productive manner sanctioned and provided by the state. New York's grandiose 1923 plan for a statewide system of parks connected by parkways promised just that: rather than merely a means to get to a state park, the ride to the park was to be a part of the experience. Scenery, whether mountainous or littoral, accompanied drivers and passengers from the moment they entered the park-parkway complex. Hundreds of miles of parkways thus appeared on the drawing boards and on the landscape of New York over the next decades. Not to be left behind, Connecticut added its own parkways.[28]

Parkways beyond the Northeast

The focus on Moses and on New York, however, obscures how broad and deep the governmentally supported movement for landscaped parkways was in the America of the 1920s and 1930s. Several states announced ambitious plans either to incorporate parkway design features into new highways and update their old ones or to build new parkway systems altogether. The highway commissioner for Minnesota announced the goal of creating a "distinctive parkway system covering the whole state" by creating funding incentives for the counties to build such roads, rather than (or in addition to) highways that would allow trucks.[29] To this day, Minneapolis boasts one of the country's longest continuous systems of public urban parkways. A loop road of fifty-two parkway miles (eighty-four kilometers) connecting various lakes and other parks called the Grand Rounds has encircled the city since the 1920s. Roads forming an arterial circle had been envisioned in the first plan for a park system, which the noted landscape architect Horace Cleveland conceived as a unity of parks and parkways in 1883, including some of the latter along the Mississippi River.[30] Initial road stretches along lakes and rivers became the Grand Rounds in the half-decade leading up to 1925 under park superintendent Theodore Wirth; paving was not completed until federal relief money made it possible during the Great Depression.[31]

The prominence of the Olmsted parkways at the beginning and the Robert Moses parkways at the apex of the American parkway movement has led some historians to overlook the landscaping of thousands of miles of ordinary highways in some American states. Roadmindedness was widespread. Before 1933, Texas and other states implemented ideas about fitting roads into their environments extensively, the significance and exposure of the Northeastern parkways notwithstanding. In addition to showing the extent to which parkway ideas circulated among and were implemented by American professionals nationwide, these developments show that highway "beautification," as it was most often called, thrived on ideas labeled "European" and sometimes proposed by European actors. The other important facet was the involvement and eventual displacement of female middle-class amateurs.

Texas, the largest continental state, was also the most interesting one in this regard: by 1940, some 9,600 miles (15,450 kilometers) of highways in the Lone Star State had been planted with trees and shrubs; plantings for controlling erosion were added to almost 14,000 miles (22,500 kilometers) of highways there. The professionals in charge of the Texas Highway Department did not merely want to add beauty to their existing highways or design new ones in a pleasing manner. Rather, beautification and conservation—in particular, erosion control after rainfall—went hand in hand. The job of promoting these arboreal implements fell to a thirty-seven-year-old landscape architect who had previously designed the urban park system for the capital city of Austin. Jacobus "Jac" Gubbels (1896–1976), an immigrant from the Netherlands with Dutch and German training, had worked in the Dutch colony of Sumatra for six years as a "plantation locater" before coming to the United States.[32] Roadmindedness traveled across borders via journals and conferences and sometimes with an individual such as him.

Gubbels, apparently, was a one-man show with considerable institutional backing. In the late 1920s, the state highway engineer had dismissed the practice of planting trees alongside highways as "European," until a persistent member of the State Highway Commission convinced him of the value of saving existing trees and scattering wildflower seeds. In 1931, Texas was the third state to set up a Bureau of Roadside Development.[33] Gubbels took over this office and used it as a platform for propaganda. Lively articles in *Landscape Architecture* carried his design proposals to a professional audience of peers. Five years into his Texas tenure, a prominent publisher issued Gubbels's book-length, accessibly written roadside treatise (*American Highways and Roadsides*), an exhortative manual highlighting the virtues of a new profession that Gubbels called the "landscape engineer." What was such a person to do?

Quite accurately, Gubbels noted that civil engineers planning a road would detest beauty "that is plastered on" and that they scorned "arty" decorations. Rather, the landscape engineer only had a place in modern planning and construction if he (!) could make the road safer, less expensive, and more beautiful, all at once. Only then would this new profession become indispensable.[34] Gubbels imagined four main tasks. First, to imagine the completed highway before construction, which depended upon understanding the engineering issues involved, and upon gaining the cooperation of civil engineers. Second, the landscape engineer was employed in creating the conditions for a safe and comfortable ride, which, third, was to occur in a beautiful way, as the landscape engineer would "clothe the completed highway in its most becoming garb."[35] Fourth, all of this would be achieved in a way that lowered construction and maintenance costs, mostly by preventing erosion.

All in all, his book was a plea for establishing landscape engineers throughout the United States rather than merely a presentation of achievements. Gubbels's account demonstrates that finding a common language for engineers and architects was the key to injecting engineering debates over highways with the kind of attention for roadsides and landscapes that landscape architects envisioned. Written in blunt, unadorned prose, the treatise presented these issues as commonsensical and economically wise, rather than culturally loaded and infused with landscape references. The focus for Gubbels—at least in his public pronouncements for a wider audience—was clearly the road, not the landscape it was in. This differentiated his approach from the writings and practices of the Northeastern landscape architects who were most happy when they could achieve parity, if not dominance over the design process, rather than seek cooperation with engineers from a subordinate position. The extensive legal and professional framework for parkways in the vicinity of New York was not the norm.

Despite these differences, all professional actors concerned with the road-

AT SLOW SPEED — 30 MILES PER HOUR — THE DRIVER'S EYE IS
FOCUSED A FEW HUNDRED FEET IN ADVANCE OF THE CAR,
AND HIS RANGE OF VISION INCLUDES THE EDGE OF THE
PAVEMENT NEAR-BY. THE ROADWAY SEEMS WIDE

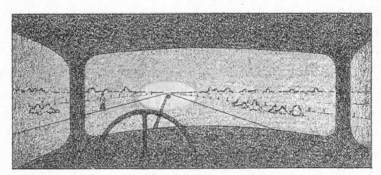

AT FAST SPEED — 70 MILES PER HOUR — THE DRIVER'S EYE IS
FOCUSED FAR AHEAD OF THE CAR, AND HIS RANGE OF VISION
IS VERY LIMITED. THE ROADWAY SEEMS NARROW

Speed and roadside perception. One of the few observers to adopt the driver's
perspective in publications, the landscape planner Jac Gubbels juxtaposes the
"driver's eye" at 30 and 70 miles per hour. At faster speeds, the driver's range
of vision is more limited, roadsides are less visible, and the road appears to be
narrow. Gubbels worked for the Texas Highway Commission and made a name
for himself nationwide with practical design advice. Jac L. Gubbels, *American
Highways and Roadsides* (Boston: Houghton Mifflin, 1938), 22

side were aided by and ultimately dismissive of the work of local women's garden clubs. In the 1920s and early 1930s, middle-class women in these clubs prominently and successfully argued for the conservation of roadside trees and the banning of billboards, with the goal of making all of Texas "one vast park." State highway engineers and landscape architects noted and at times amplified these female voices, as they lent prominence and middle-class respectability to their agenda. When the time came, however, to voice concerns over roadside issues within a professional structure, with offices, jobs, and regulative power, state officials predictably deemed this issue to be a "man's job."[36]

Gubbels was one of the few writers among landscape architects to pay attention to the sightlines of motorists traveling at higher speeds. His book considered landscape as viewed through a windshield. Compared to other authors on roadside improvement in the pages of *Landscape Architecture* during these years, Gubbels is one of the most lively and eloquent. But he was not an outlier. The Bureau of Public Roads employed Wilbur H. Simonson (1897–1989), a prolific planner and writer, from 1929 until 1965.[37] He became chief of the Bureau's "Roadside Branch" from 1932 until his retirement. By the 1940s, it had become common practice for state highway departments to hire landscape architects. Again, this did not mean they had decisive influence on planning. In the pages of their professional journal, landscape architects often assumed a defensive posture by asserting that they were not "pansy-planters"; thus, they asked to be involved before and during the highway planning phase, not afterward.[38]

Simonson's showpiece was the Mount Vernon Memorial Highway, the other parkway to spread the gospel of roadmindedness widely, both within the United States and abroad. Built to commemorate the two hundredth birthday of America's first president in 1932, the federal government sponsored the construction of this road leading from the nation's capital to George Washington's plantation. On its approach from the city, motorists experienced the closing and opening-up views of the Potomac River. Remarkably, the otherwise utilitarian Bureau of Public Roads employed the same landscape architects who had worked in Westchester County for this special occasion. This road marked the entry of the Bureau into the realm of parkways. Bureau chief Thomas MacDonald sought "as close an approach to nature as can be managed."[39] In contrast to the restoration goals of the Bronx River

Parkway, the Mount Vernon road treated the Potomac river as mere scenery, not an environment to be rebuilt. In fact, the highway plans conflicted with efforts to turn both banks of the river from Mount Vernon to Great Falls into a preservation area. The road builders' techno-natural argument prevailed.[40]

American parkways defined landscapes for urban motorists and made them the vital ingredient of the simultaneously individualistic and prepackaged scenic driving experience. Bringing nature closer to city dwellers was celebrated as a democratic achievement and thus a token of Americanness. In a historically rare coalition of professional groups, landscape architects and civil engineers presented parkways as a progressive means to egalitarian consumerism that would mend the rupture between country and city. The Mount Vernon Memorial Highway was to be as patriotic as the Washington Monument and as modern as the newest earthbound mode of circulation and mobility. While urban and suburban parkways in other locations did not achieve the fame of the Mount Vernon road, they brought this vision of blending nature and technology to millions of Americans.

The Allure of Fordism

The Bronx River Parkway and the Mount Vernon Memorial Highway were only two of the suburban parkways constructed during the interwar period. Yet, seen from the vantage point of the international circulation of knowledge and their contribution to roadmindedness, they stand out. For the first half of the twentieth century, Americans owned more automobiles than the rest of the world combined. This made cars and roads in the United States an attractive object of study from abroad. Manufacturers, planners, and government officials were more than happy to receive this growing attention. During the interwar period, "the Germans made a second discovery" of the America of Edison and Ford, writes historian Thomas P. Hughes.[41] Touring the massive River Rouge Ford factory in Michigan became obligatory for engineering and other experts visiting the United States. Fascination with the output of this plant and astonishment over the extent of consumerism were some of the characteristics that garnered their interest. The prominence of automobiles was not lost on any German observer. Experts in civil engineering and landscape architecture immersed themselves in the study of American parkways.

The fact that relatively few Germans owned automobiles, especially in comparison to the United States, made such visits to and reports from the land of mass motorization more appealing, not less so. Promoters of cars and roads were eager to link technological modernity with the rise of the automotive sector, and to posit America as a vision of what was to come in Germany. Together with aviation, automobility possessed a futuristic cachet, which contributed to roadmindedness.

During the interwar period, a German lobby presented a vision of long-distance roads. Hamburg, Frankfurt, and Basel were to be connected with the acronymic Hafraba Road. The inspiration came from Italy, where the *autostrada* had received Mussolini's support. Close to three hundred miles (five hundred kilometers) of toll roads for cars and trucks were built by the mid-1930s. The Fascist regime fancied the infrastructural prowess of new roads, but it paid considerably less attention to landscaping than either German or American planners.[42] North of the Alps, efforts to construct a national highway network went nowhere during the years of the Weimar Republic. Undeterred, landscape architects published papers and plans on how to design the roads of the future together with their surroundings. Many referenced American parkways as examples of scenic highways.[43] The idea of parkways as automotive corridors in parks, however, appeared alien to them. Urban public parks tended to be much smaller if they were recent, and cars would have taken up spaces for pedestrians. Older parks, often the remnants of feudal hunting or pleasure grounds, tended to be larger, but bourgeois aesthetic tastes for walks favored a retreat from hectic modern life, of which automobiles were only the most recent symbol.[44]

Before the Nazi takeover, German states and provinces spent some money to adapt existing roads for automotive traffic by widening and paving them. New construction, especially on a large scale, was a subject of learned debates, based on the expectation that car ownership would expand in the future, rather than on an assessment of reality on the ground. Engineers expanded their knowledge during this decade by researching different types of base layering and surfacing. They standardized and published extensively. As far as car-only roads were concerned, a stand-alone twelve-mile (twenty-kilometer) stretch of highway between Cologne and Bonn opened in 1932, but it remained an anomaly. Automobiles were sparse, especially outside of affluent urban areas, and the railroad network was extensive and successful, both for freight and passenger transport. While mass consumerism was on the rise in Germany, cars were far from being affordable consumer items.

Pointedly, the economist Werner Sombart dismissed car-only highways as "roads for the cheerful enjoyment of life by the rich."[45] The state-owned railroad ruled transportation. In fact, it provided steady income for the Berlin government. Supporting the competitor of rail transport made little political sense, therefore. This was the opposite of the situation in the United States. American privately owned railroads had been involved in corruption, were accused of trying to monopolize traffic, and farmers were eager to compete with what they saw as corporate behemoths by using their own vehicles on farm-to-market roads.[46]

Autobahns and the Nazi Dictatorship

The most prominent Continental European expression of roadmindedness, the Nazi autobahn network, was supposed to represent the Nazi regime's dictatorial embrace of power. In contrast to the virtual absence of road construction during the Weimar Republic, and despite the lack of automobiles on these roads, the regime pushed for the planning and construction of a 2,500-mile (4,000-kilometer) network. Seifert and other landscape architects who were not defined as Jewish embraced the infrastructural frenzy of the early Nazi years as a professional boon.

Quite literally, the autobahn had Hitler's name written all over it. While it would be inaccurate to adopt the propaganda claim that the autobahn was "Adolf Hitler's roads," it would be equally misleading to discount the dictator's impetus and influence. He embraced the highway network as a propaganda tool and as a token of Nazi modernity. Germany's jump-start into individual motorization in the 1930s, although haphazard, was economically, politically, and culturally significant. Economically, the dictatorship aimed to transform the transportation sector by adding a layer of cars and roads to the country's extensive (and dominant) public transportation system; politically, the push toward motorization served the goal of portraying the fledgling regime as activist and goal-oriented; and culturally, the car-road complex was imbued with an aura of novelty and modernity. As one celebratory newspaper put it, Hitler had done away with the "medieval opposition to the automobile."[47]

In the spring of 1933, political enemies of the regime were rounded up and sent to recently established concentration camps; Germans defined as Jewish were legally excluded from holding positions in public administration, including universities; and Hitler gave a speech envisioning Germany as a country crisscrossed by four-lane highways populated by a racially pure populace riding in their own automobiles.[48] This was a regime that prized constant activism and frenzied efforts at mobilization in every respect. Thus, putting more people (or men, to be more accurate) behind steering wheels on new roads was central to the regime's efforts and self-portrayal. Moving troops and military equipment were not the main goals for investing in these roads; in fact, German generals opposed them in closed meetings. Automotive leisure was to be the result of these efforts: "Weekend, *Kraft durch Freude* [the regime's leisure agency], Volkswagen—all three serve the great Nordic inclination to overcome the tightness of territory at least for recreation," as the highway official Fritz Todt averred.[49]

Cars and roads were part of a Nazi effort for a racially defined consumer society. Inspired by Fordist mass production, and the writings of Henry Ford himself, Hitler embraced these conveyances as harbingers of modernity. Ford's antisemitism also did not hurt, to be sure. The German terms for the celebrated cars and roads mattered: as *Volkswagen*, people's cars, the automobiles were to be popular and affordable, bringing car ownership to the masses; and the *Reichsautobahnen* signaled the central state's deep involvement in roads. Usually, cars were driven locally, on weekend outings, or for road trips on shared roads. The new highways, however, were meant exclusively for cars and trucks, to be driven at constantly high speeds, without having to worry about intersections, oncoming traffic, pedestrians, bicyclists, or any other mode of transportation.

With these efforts, the Nazis embraced a distinct form of consumerism, using cars and roads to glorify the state and its powerful leader. At the same time, consumerism was individual. Volkswagens were supposed to be affordable with a savings plan. Historians have been debating whether the latter effort was a genuine foray into consumerism or merely the attractive facade behind which the regime hid its more reprehensible policies of categorization, exclusion, and extermination. It is clear, though, that the specific Nazi mode of consumerism was intended only for citizens defined as Aryan. The automobiles of many middle-class Germans classified as Jews became the loot of Nazi thugs early on. In 1938, the regime limited their mobility and stripped them of driver's licenses.[50] Defiantly, the Dresden professor Victor Klemperer held on to his car as long as he could and used it to temporarily escape the increasing pressures of being defined as Jewish. His diary entries speak to these escapist moments.[51]

The Nazi autobahn with Dachau exit sign. This image of an empty autobahn stretch speaks to the paucity of automobiles in a country that built thousands of miles of interstate highways. The politics of automobility were contradictory in Nazi Germany. While the highways materialized, cars for the masses did not. Consumerism during this dictatorship was racialized and based on political acquiescence. This stretch of the highway features an exit ramp for the city of Dachau, where the regime's first concentration camp was located. Staatsarchiv Munich, Autobahndirektion Südbayern, Holzkasten 1, 20020

In contrast to the bombastic propaganda, the motorization effort in the form of the Volkswagen failed dismally during the Nazi years. For German car manufacturers, an affordable car for the masses made little business sense, given their profit margins on large vehicles for affluent consumers and the lower purchasing power of the working class. In response, the dictatorship set up a state-run company built from scratch. A savings plan, popular among racially selected consumers, created expectations for a German version of the Model T. But the production facility, built in Northern Germany and modeled after the River Rouge facility in Michigan, mostly produced vehicles for the Reich's war effort.[52] Incongruously, some 2,400 miles (3,800 kilometers) of autobahn were built in Germany and Austria before the beginning of World War II shifted priorities for the Nazi regime. The autobahns were meant to impress not only with their scale—no country at the time could boast such a network—but also with their quick construction. With only slight exaggeration, a propaganda movie claimed that three kilometers of roads were built per day during the peak of the construction frenzy.[53] The net result was that the speed of constructing the autobahn outpaced the growth of the car sector by far.

The promise of a popular car and the effort put into planning and building the autobahn network were magnified by Nazi propaganda. The scale of the highway project and its rapid construction were celebrated over and over again. The regime unleashed a torrent of books, magazines, movies, newsreels, theater plays, and board games to tout its version of roadmindedness. Every kind and level of media was involved. Portrayals of movement and circulation were staples of Nazi parlance. The propaganda exaggerated the modest effects of road construction on unemployment relief. It claimed that the autobahn network would not only provide transportation benefits and stimulate the economy, but also enhance, rather than despoil, the landscape. The dictatorship commissioned painters and photographers to convey this notion pictorially.

Consumers of this multimedia onslaught would not have been able to ascertain the international underpinnings of such claims. Building a scenic highway rather than merely a utilitarian one was a specifically German and Nazi achievement, according to the regime's top engineer, Fritz Todt (1891–1942). He supervised the autobahn project from 1933 onward and affirmed at one of the widely broadcast Nuremberg party rallies that "the National Socialist road builder prizes the cultural and landscape value of his new roads at least as highly as the purpose of material transportation. The National Socialist loves his homeland not only through word and song, but through deeds."[54] In Todt's and the regime's public pronouncements, the roads were German through and through. Yet, when Todt commissioned Seifert to offer his first landscaping advice for the autobahn, the engineer introduced the notion of a "Parkstraße," obviously referring to American parkways.[55]

Landscape architects and civil engineers paid close attention to the development of parkways in the United States; yet in public they stressed that only Germany was capable of presenting its landscapes in such a modern, motorized version. The hypernationalistic Nazi regime claimed that its road network was homegrown, but the expertise used in planning and building these roads was international. Todt himself was a keen observer of the American roadbuilding scene. He worked for a construction company in Munich during the interwar period, planned local and regional roads, and followed developments abroad, including those of the United States—as any alert civil engineer of his generation would. At the same time, Todt was an early and enthusiastic member of the Nazi party.[56] During the dictatorship, Hitler entrusted Todt with the autobahn project, infrastructure planning, and industrial war production.

Given the rapid pace of autobahn planning and construction, and a paucity of civil engineers elsewhere, Todt recruited those from the country's railroad and put them to work on its roads. The professional training and experience of these engineers would thus guide them to design roads along the lines of railroads: as straight and unadorned as possible. Such ideas flew in the face of wanting to build "Parkstraßen," or parkway-like roads. When Seifert offered his services to Todt, the latter appointed him as an advisor and asked him to select a cadre of landscape architects to consult with the engineers. They were to act as counterweights to the railroad engineers. Seifert, whose freelance work had dried up during the Great Depression, cheerfully obliged.

The results of these efforts were rather mixed, however. Unlike in the United States, trucks shared the road with cars. More importantly, landscape architects and civil engineers clashed rather than cooperated; the scenic qualities of the autobahn were hotly contested and only sporadically realized.[57] The first stretches of the autobahn resembled railroads and made for monotonous driving. This wouldn't stop the propaganda from claiming that they were fully immersed in the landscape. Later extensions did direct drivers to vantage points, especially in proximity to the Central German Uplands and the northern rim of the Alps. The goal, according to Todt, was to give the highways a "scenic character immanent in German essence."[58] While the propaganda would not tire of touting these assumed qualities, the planning and construction mirrored the muddled and contradictory political character immanent in Nazi governance. Established institutions and actors would be set to compete with new forces and ideas. These battles were overshadowed by bombastic claims and inconsistent execution.

This German version of roadmindedness caught the attention of foreign observers. A Pennsylvania newspaper, while impressed with the scope of the project, noted that a "small army of landscape architects" was working on the "horticultural decoration of the speedway"—a job description the architects would have abhorred, since they aimed to deeply embed the road in the

A fast ride on the autobahn. While using some of the rhetoric of embellishing the landscape rather than despoiling it, the planning and construction of the autobahn did not live up to the standards proclaimed by its builders. Many of its stretches, especially the ones built early on, were designed with railroad parameters, such as this straight and level section allowing for a fast ride through a forest. Staatsarchiv Munich, Autobahndirektion Südbayern, Holzkasten 3, 20157

land, not merely add floral icing. At any rate, the American paper realized that the goal was to "give the maximum of beauty plus utility, but some sections have been laid out with no other object than to provide the traveler with aesthetic satisfaction. Chief of these is the Alpine Highway."[59] Gilmore Clarke provided a less flattering assessment of the autobahn, deeming it of lower aesthetic quality than American parkways.[60]

For Seifert, the autobahn project and the Alpine Highway replaced the professional uncertainty of the Weimar Republic. Garden architects had mostly relied on commissions by private homeowners. Large public parks were rarely planned at the time, and infrastructural projects languished during the Weimar period. Due to the relative paucity of automobiles in Germany, calls for building new roads on a large scale were the province of technological enthusiasts and utopians. Seifert had not been one of them, but he benefited from the torrent of infrastructural activism unleashed by the Nazi regime after 1933. "Scaling up" could have been his motto for the dictatorial years, as he made the transition from designing private backyards to planning a nationwide highway network; as his profession increasingly substituted entire landscapes for private gardens, his political acumen, unfettered by scruples, catapulted him from the position of a provincial architect to the job of advising an expansive dictatorship.

Designs and Politics

Design professionals such as Clarke and Seifert helped to give meaning to automobility in the interwar period. Both pursued remarkably similar visions of roadmindedness in drastically different political regimes. Clarke relied on the patronage of Robert Moses in New York and its environs and used this position to become the leading advocate for parkways in the United States. But it is also important to remember that parkways and landscaping highways thrived elsewhere, too, as the examples of Minneapolis and Texas show. By contrast, German landscape architects such as Seifert observed the American plans and highways. Roadbuilding on the scale of the United States was inconceivable in interwar Germany, low as the number of automobiles was. But parkways offered a glimpse of how landscape architects could broaden their professional portfolios and become involved in infrastructural work on a larger scale. The Nazi dictatorship, with its sudden and emphatic embrace of cars and roads, provided just that. Seifert's patron was Todt, whose visions of German roads were inspired by American highways and parkways.

Driving or being driven served goals of national politics and would compensate for environmental losses. As we will see, Seifert's ideas of nationhood were more exclusionist than Clarke's vision of an ordered movement guided by professionals whose status was uncontested. However, they were united by the idea that automobility could cure some of the ills of modernity, if it was guided by experts such as them. These scenic infrastructures could be molded and remodeled. Parkway designers were not aiming to reestablish a pre-modern, pre-automobile, or even pre-railroad landscape. Rather, they wanted to make sure that the newest terrestrial transportation technologies of the day, cars and roads, would be controlled by professionals sensitive to the repercussions of the rise of automobility. While the proto-ecological restoration of the Bronx River and its valley in the vicinity of the

June 1938 cover of *Fortune* magazine by Hans Barschel. This issue featured a ten-page article on Robert Moses and his infrastructural activities. The magazine's cover shows an abstracted tangle of roads superimposed on a forest with bridges, interchanges, a tunnel, and roads reaching ever further, seen from a bird's-eye view. The new scale and scope of roadways was meant to impress in and of itself, without a single automobile on these roads. They appear sculpted and dynamic. The graphic designer for the cover was Hans Barschel, who, just a few years earlier, had designed the poster for a large 1936 automobile exhibit in his native Germany. After emigrating to New York in 1937, Barschel began to work for American customers. *Fortune* © 1938 Fortune Media IP Limited. All rights reserved. Used under license. Fortune and Fortune Media IP Limited are not affiliated with, and do not endorse products or services of, University of Maryland.

Bronx River Parkway might seem like an effort to return the area to its pre-nineteenth-century condition, it would be more accurate to understand its result as an expertly managed modern landscape that sought to integrate a contemporary roadway for unencumbered, aesthetically pleasing movement of motorists, and a rejuvenation of a heavily used landscape. Neither man nor nature was paramount in such an environment—the effortlessly moving automobile was.

In terms of architectural styles, the regionalist architecture of the parkways put them somewhat at odds with modernists. Bauhaus architecture and the Modernist movement in general were anathema to both Seifert and Clarke; the former was glad to see the Bauhaus leave Nazi Germany, while the latter was incensed by its prominent role in exile at the Harvard Graduate School of Design. Having served on the National Commission of Fine Arts from the 1930s to the 1950s, Clarke detested but could not prevent modernist architecture on the National Mall in Washington. He was particularly indignant at the building of the Hirshhorn Museum, a concrete cylinder on legs, designed by the modernist architect Gordon Bunshaft, which opened in 1974, and the East Building addition to the National Gallery of Art, which opened in 1978. While the former was "totally out of character" with the rest of the Mall, according to the grumbling octogenarian Clarke, the latter was "even more hideous" and, to boot, designed by "a Jap named I. M. Pei."[61] It would be pedantic to point out that Pei was, in fact, Chinese American; what really matters is that for Clarke, architectural modernism and Pei were equally foreign to one of the nation's politically most significant public spaces.

Scenic infrastructures were contested spaces in the twentieth century. Roads and their landscapes were never mere transportation corridors. The ideology of roadmindedness had helped to elevate roads to levels of cultural, social, and political significance that they had rarely enjoyed before. They had become desirable elements of modernity, yet their relationship with architectural modernism was fraught. The emphasis on roadsides and landscapes opened up the realm of environmental remediation through mobility.

Roadmindedness was based on international exchanges, visits, and publications. By the 1920s, the United States saw itself as the lodestar of the automotive universe, what with its millions of mass-produced vehicles and thousands of miles of new roads. Among the latter, parkways took on a special meaning for Americans and for visitors, as these highways aimed to provide more than just movement. For a country such as Germany, with its relative scarcity of automobiles and no new roads to speak of, they became a reference point. Some of their design features entered the lexicon of auto-bahn planners, even though the results were inconclusive.

Both American parkways and the German autobahn were landscapes of exclusion, although at different levels and to different degrees. More so than other such roads, the Bronx River Parkway erased immigrants and their houses from the landscape to create a pastoral view seen from the scenic infrastructure. Not just on the autobahn but on every road, Germans de-fined as Jewish were legally banned from driving by 1938, thus ruling out this form of movement for them.

Even though they were predicated on the cross-border, intense exchange of knowledge, scenic infrastructures became increasingly inward looking. Parkway planners, especially Clarke's group, emphasized the novelty and originality of their approaches. Germany's Nazi dictatorship unsurprisingly disavowed any foreign parentage for its nationalistically charged projects. But the parallel histories continued. By the 1930s, the governments of these two countries were to sponsor even more ambitious efforts to reunite mo-torists with their landscapes.

3

Roads in Place

Otto Klenk was happy. Out for a Sunday drive, the middle-aged Bavarian enjoyed the landscape just outside of Munich. Yellow and green colors dominated. After passing through a thick forest, Klenk relaxed, a pipe dangling from his mouth. Having lowered his speed, he contemplated how well meadows, forests, lakes, and mountains were composed as he gazed upon them through his windshield. Klenk looked forward to a meal with his paramour and some hunting.

All of a sudden, a comfortable-looking touring car passed him. Both driver and passenger appeared strange and exotic to Klenk, the native Bavarian. His mind began to wander. More people were on the move, he recalled having read, which would lead to a new mass migration: slow and sedentary people would be pushed aside by agile, nomadic types. Not amused by the prospect, Klenk stopped his car and looked at a roadside memorial for a farmer killed as he transported hay. Another car stopped. To his great chagrin, Klenk heard Northern German voices trying to read the Bavarian inscriptions. The strangers, it would appear, were already here.

Increasingly tense, he drove off and accelerated, no longer enjoying the scenery. Klenk's thoughts turned to work-related matters. When not maneuvering a vehicle, Klenk ran the Bavarian justice department. Adversaries such as the liberal lawyer Siegbert Geyer—a "dirty Jew" to boot—bothered him. But the pleasant, multicolored landscape calmed Klenk down. He thought of his conservative fellow party members and how simpleminded some of them were. Klenk's car almost hit a bicyclist, which led to an exchange of tirades.[1]

While mountains, lakes, anti-Semitism, roads, and automobiles all existed in Bavaria in the 1920s, Otto Klenk did not. He sprang from the pen of the writer Lion Feuchtwanger, who painted a lively tableau of Bavaria in his 1930 novel *Success*. In the book, the justice minister spends most of his professional energy hounding an insubordinate museum director. Conservative politicians, supportive clergy, and self-satisfied Bavarians populate Feuchtwanger's book. Some of the fictitious characters bear traces of historical figures: a likeness of Adolf Hitler draws crowds in the beer halls; a fashionably ragged poet with a fondness for automobiles resembles the young Bertolt Brecht.[2]

Klenk's sudden shifts of attention from scenery to politics and back presaged the intimate relationship between the two in the 1930s. Driving a car and seeing one's surroundings, while never merely an innocent, personal act, became dramatically and overtly politicized in the 1930s. At the political level, the United States and Germany drifted apart. One remained a democracy, the other one turned into a violent dictatorship by 1933. But both made cars, driving, and the consumption of groomed autoscapes tokens of national belonging. Roadmindedness reached new levels, found powerful patrons, and resulted in extensive infrastructures. Automotive landscapes not only received the financial and administrative support of central governments; they became elements for educating the citizenry (differently defined, of course), and of ordering and displaying landscapes. The attraction of a car ride immersed in scenery, as fleeting a beauty as there ever was, was meant to contribute to a deep and permanent sense of belonging.

For decades, the reality, if not the ideology, of the parkway had been about exclusion—in the sense of excluding people without automobiles, inappropriate signage, and undesirable views. Yet, when central governments stepped in and adopted parkways on a national scale, the terms of exclusion and the meaning of these parkways changed substantially. As scenic roads became national, they grew in size and importance. They still showcased particular versions and visions of regionally understood landscapes. But their designers aspired to create new regimes of space and time. Nationally funded and orchestrated parkways were to be symbolic of modern nation-states and their increasing reach into the farthest-flung areas. As such, parkways marked territoriality. These spatial arrangements also rearranged time. While undoubtedly modern, national parkways mobilized landscapes and narratives of rural pasts integrated into twentieth-century environments of belonging and consumption.

The two most extensive parkway efforts in these two countries were the Blue Ridge Parkway in Virginia and North Carolina, and the German Alpine Road (Deutsche Alpenstraße) in Bavaria. As the purposely designed nexus between technology and landscape, they were showcases of a mobility-oriented rescaling of the driving experience. Centrally planned and often opposed by locals, these landscaped roads were grand monuments to a new form of driving and tourism. A particular form of roadmindedness emerged. While presenting locally and regionally understood landscapes, their origins, design, and management were metropolitan rather than rural. Their smooth and pleasant appearance belied conflicts, beginning with basic routing issues and not ending with signage. They were duplicitous landscapes, as the geographer Stephen Daniels would insist: their attractive features glossed over conflict, contestation, and context.[3] Harmony, not disagreement, continues to be the basic message of these ensembles, whether seen while driving or in photographs. Upon closer examination, disagreement, not harmony, is the basic message of their history. Today, both harmony and disagreement are so entangled as to form a mangled unity.

Roads in the Great Depression

As public works projects, these two scenic roads emanated from crises and contingency. The Blue Ridge Parkway and the German Alpine Road had been proposed by locals but only became a reality when the Great Depression demanded visible efforts to overcome economic calamities and put people to work. The politics of the 1930s—the New Deal and Nazi economic efforts—made these roads, their extent, and amounts of funding possible. As large-scale infrastructures supervised and financed by national governments, the two parkways reflected the political systems of which they were a part: a more interventionist democracy and a centralizing dictatorship.

In the United States, New Deal policies sought to transform both the economy and the environment. The economic crisis of the early 1930s provoked stronger state action; transformationist visions abounded.[4] Among Washington planners, it was common wisdom that better management was needed for the economy and the environment. A stronger central state supplied the institutions and the financing, millions of men provided their labor in the Civilian Conservation Corps (CCC), the Works Progress Administration (WPA), the Tennessee Valley Authority (TVA), and elsewhere; as a result, the New Deal period "perhaps more than any other in U.S. history witnessed the transformation of public space by the federal government."[5] More and more rivers, forests, and agricultural lands saw at least an effort by Washington, DC, to add a layer of federal intervention, supervision, or management. The Soil Conservation Service sought to convince farmers to implement different farming methods to prevent another Dust Bowl, the catastrophe that had ravaged the Great Plains. Some landscapes, such as millions of acres of what became state parks, changed owners and purpose. Through the work of CCC men, pastures and commercial forests became landscapes of recreation. The Blue Ridge Parkway is one of these landscapes that owe their existence to the New Deal.[6]

Mobilizing landscapes and people was one of the hallmarks of Nazi Germany on its way to World War II, albeit under different auspices than in the United States. The economy recovered after the Great Depression; the regime began to prepare for a war. Unlike its American counterpart, the German economy could not rely on domestic resources and food sources alone. Under the banner of autarky, abandoned coal and iron ore mines reopened, production accelerated at existing ones, and factories and cities grew. The German counterpart to the Civilian Conservation Corps, the Reich Labor Service, worked rarely on recreational landscapes but cleared forests for roads and drained swamps, thus bringing marginal landscapes into the reach of agriculture and human settlement. Simultaneously, Hitler's regime unfolded a repertoire of infrastructural activities. Expanding transmission lines for electricity, constructing hydroelectric dams, and, not least, roadbuilding served the goals of preparing for a war of aggression, boosting industrial production, and tying a racially defined citizenry closer to its Führer.[7]

The autobahn, as noted, was the most prominent of these projects. Its landscapes were hardly the unalloyed boon to the environment that the regime made them out to be, although a green sheen characterized the early history of the regime in general. German conservationists had placed high hopes in the new regime. In its early years, the dictatorship displayed somewhat surprising degrees of affection for natural spaces, at least on a rhetorical level. Environmentalists and some of the members of the Nazi elite became temporary bedfellows. The first nationwide conservation law was a product of this union. Hundreds of nature preserves were cataloged. Yet preservationist concerns tended to take a back seat to the goal of preparing for war. In simplified terms, Nazi environmentalism was shallow but wide.[8]

Seen in this light, the regime's roadways are test cases for its proclaimed environmentalism and its transformative approach to landscapes. The German Alpine Road was designed as a scenic showpiece: the Alpine and subalpine topography of Southern Bavaria was put on a pedestal of concrete and asphalt. Both the dictatorship's ostensive love of nature and the Nazi variant of roadmindedness came to the fore. While a few locals had suggested similar scenic infrastructures, the Alpine Road was a thoroughly urban and metropolitan project.

In comparison, the planning and construction of the Blue Ridge Parkway reflected the New Deal's infrastructural activism and its sponsorship of car-based tourism in a more democratic system, yet with similar tensions between planners and locals. These altercations played out against the background of remoteness: Southern Bavaria and Appalachia were both regions most easily identified with social and cultural backwardness, agricultural economies, poverty, and lesser degrees of political representation. In addition, tourism emerged as an economic strategy and cultural marker for both places in the twentieth century.

Alpine Road Proposals before 1933

The Blue Ridge Parkway and the German Alpine Road were both elevated roads: as the most extensive scenic roads in their respective countries, they received large amounts of money and attention. They were also mountain roads, designed to capitalize on viewing opportunities from high above and to present a motorized version of mountainous communities. Although the Appalachians and Alps are quite different topographically, the approach of the designers to the heights was similar: they turned mountainous landscapes into consumable scenery.

The most obvious difference between the two mountain ranges is their height. The highest Alpine summits, situated above the timberline and covered in perpetual ice and snow, appear forbidding. The tallest peaks reach more than 13,000 feet (4,000 meters). Yet, Alpine landscapes were not devoid of humans. Permanent settlements in the valleys were long-established. As one historian put it, the Alps acted "more as a filter than a barrier" for human history, including trade and migration. Forests provided building materials and firewood for centuries, mines were common, and pastoralists built livelihoods around the environment and the seasons. During the summer, Alpine pastures became sites of transhumance. Cow's milk became more portable and lasted longer once it was turned into cheese, using human labor and assistance from enzymes in the form of rennet. Geological forces had shaped the mountains; humans and other animals sought to reshape them for agriculture and trade.[9]

In comparison, the summits of Southern Appalachia feature less extreme heights, but a series of mountain ranges of 6,000 feet (1,800 meters) and more. Native American interventions on the landscape appear light in comparison to agriculture and silviculture practiced by European settlers and their descendants. These transformations remained mostly local in scale before the nineteenth century. Timber and mining interests, and the growing involvement of the region in capitalist exchanges, remade the ridges, slopes, and valleys more systematically from then onward. By the early twentieth century, drastic deforestation marked several areas. With accelerated timber cutting came railroads and mills. Enough scenery remained, however, to support a nascent tourist industry.[10]

Bavaria and Appalachia both came to be viewed as exotic as they became tourist regions.[11] Within Germany, cultural differences and regional allegiances have been so pronounced as to be stereotypical. The twin forces of political unification in a new Germany after 1871 and of economic integration, both domestically and globally, helped to contribute to regional senses of belonging wrapped in national outlooks. Rhinelanders, Saxons, and Swabians posited themselves as such first and as Germans second. Food, landscapes, and dress made the difference between one region and the next. Among these quasi-tribal identities, Bavaria's sense of self provided a counterpoint to Prussian hegemony and the powerhouses of industry within Prussian borders. While Munich developed into a fair-sized city with a sizable middle class, and Northern Bavaria saw pockets of industrialization and relative wealth, Southern Bavaria apart from Munich and Augsburg remained overwhelmingly rural. The plains, with their fertile soil, supported farming communities and small towns, while the more mountainous valleys and higher elevations allowed marginal agriculture and silviculture.[12]

In the larger contexts of the European Alps, the mountains on Bavarian soil were neither particularly high nor particularly distinctive. The retreat of glaciers after the Late Pleistocene left a legacy of valleys and lakes, as it did elsewhere. Switzerland and Austria offered more summits at higher altitudes, and the former led Europe in establishing mountain-based tourism. Bavaria was varied enough, topographically speaking, to feature a few high points, but it contained mostly foothills and plateaus within its borders.

However, seen from the vantage point of Germany at large, the Bavarian Alps stood in sharp contrast to the large North German plains and the Central German Uplands. Bavaria was the only German state with access to this mountain range. Seen from Berlin, the center of national political power, the Alps were the periphery of an economically middling and partially poor Southern state.

Still, Bavaria's Alps mattered greatly in the cultural imagination. While other regions provided their share of imagined backward-oriented groups struggling in the face of modernity, the image of the eternal Bavarian mountain-dweller dwarfed by peaks and accompanied by livestock resonated widely within Germany. Even though Alpine farmers rarely wore them, lederhosen and dirndl became synonymous with Southern Bavarian dress, due to the work of regionalist, folklorist clubs that focused on popularizing imagined or real peasant garb. At the same time, Alpine agriculture became increasingly integrated into national economies, with milk and milk products being some of the most obvious commodities.[13]

In other words, this was the ideal environment for urbanites to imagine as tranquil and unspoiled, even if humans had remade it for centuries. The result was a mental landscape of primitivism and a picture of a physical landscape of beauty, both ready for tourism. In addition, what made Southern Bavaria distinct was the infrastructural activity of a nineteenth-century admirer of this region: Ludwig II, the eccentric Bavarian king, who died in 1886. He was the consummate castle builder. Ludwig's building frenzy fed on a political desire to imprint his legacy on the Bavarian landscape in an absolutist fashion while being bound as a constitutional monarch. Aesthetically speaking, his castles reimagined a medieval and feudal past; some buildings were to create ideal vessels for the operas of his protégé, the composer Richard Wagner. More prosaically, cabinet members and state administrators were aghast at the runaway costs of construction and the resulting millions of Goldmarks of public debt. To mitigate the expenses, administrators sometimes even prescribed faux marble instead of real stone.

After Ludwig's demise, the state took its revenge. By opening the fantasy castles to the public, Bavaria more than recouped its missing millions. Less than six weeks after the king's death, the Neuschwanstein, Linderhof, and Herrenchiemsee citadels were accessible to anyone able to pay the entrance fee.[14] Immediately, guidebooks began to perpetuate stories about the "mad" king, interwoven with appreciation of the art and architecture of his build-

Sally Israel with three acquaintances in Bavarian country costume, Bad Reichenhall, ca. 1920. While on vacation, urban tourists pose for photographs in lederhosen and dirndl in a photographer's studio against a painted background of Alpine peaks and a valley. Studios would produce postcards of these scenes that tourists could send to friends and family. While not authentic, such costumes were popular and became part of local lore. By propagating such clichés and by making the foothills of the Alps accessible, Southern Bavaria developed a tourism industry that relied heavily on urban travelers such as this group from Berlin. The spa town of Bad Reichenhall, which was more accommodating to Jewish tourists than other locations, drew the ire of anti-Semites during the interwar period. Jewish Museum Berlin, Inv.-Nr. 2005/136/19, donation of Monica Peiser ings.[15] Since then, Neuschwanstein alone has attracted many millions of visitors.[16] In the process, it created a visual legacy reaching all the way to Disneyland: reproduced as "Sleeping Beauty Castle," it has graced the Southern California landscape since 1955.[17] With only slight exaggeration, journalists have claimed that Neuschwanstein, "the 'authentic' Disneyland," is as famous as the dome of St. Peter's Basilica and the Egyptian pyramids.[18]

Also distinguishing Southern Bavarian tourism are the Oberammergau passion plays, which had existed as local religious rites since the early seventeenth century. They gained prominence and visitation numbers by the late nineteenth century, when easier transportation and greater wealth helped turn them into an international spectacle. Thomas Cook, the British travel agency, organized package tours for British and American visitors as early as 1880.[19] The combination of a performance with old Catholic roots and ease of access via modern transportation proved successful, despite discussions over the anti-Semitic aspects of the plays. Almost half a million visitors attended the 1922 performances; among the visitors at a 1934 performance was Adolf Hitler, for whom the passion plays amounted to an exercise in theatrical anti-Semitism.[20]

Castles and passion plays, mountains and lakes, and hiking and climbing marked the history of tourism in these places. Locals in lederhosen performed for visitors whose access to the sites depended on rails and roads. The mountains themselves saw a growing web of hiking paths, signposts, and huts provided by the thriving Alpine clubs. In the interwar period, tourism spots such as Füssen or Berchtesgaden embarked on a strategy of expansion. Tourism grew in economic importance, especially in rural regions; by the early 1930s, more than three out of four residents in the town of Garmisch lived off of it.[21]

Tourism managers, while reassured by the proximity of Southern German cities such as Munich and Augsburg, where disposable incomes grew, constantly worried about the possibility of tourists moving on to Austria or Switzerland. Bavarian mountains were not as high as the Swiss ones, its railroads were not as accommodating to sedentary tourists, and car-based tourists found Austrian and Swiss roads to be more exciting. Travel advocates responded with motorized tourist thrills, such as the cog railroad to Germany's highest peak, the Zugspitze (9,718 feet or 2,962 meters above sea level), which was in operation by 1930. In addition to rails, building roads was another effort to prevent tourists from venturing into Austria. For example, tourists to Neuschwanstein numbered in the tens of thousands by the mid-1920s, but the shortest way from that castle to the nearby Linderhof palace led through Austria. Tourism promoters in the nearby town of Füssen suggested a new road to Linderhof on German territory to keep visitors from straying, supported by a motoring club. Preliminary engineering studies came to naught, but competitive roadmindedness persisted.[22]

The most extensive road proposal came from a locale close to another one of Ludwig II's castles. On the largest island of Lake Chiemsee, an updated imitation of Versailles called Herrenchiemsee attracted tourists. The chief tourism promoter in the lakeside town of Prien popularized the idea of a new road, reaching from Bavaria's easternmost town, Berchtesgaden, not far from Salzburg, to its westernmost large community in Lindau, on Lake Constance. The road was not to leave Bavaria for its entire length. The Prien promoter, August Knorz (1876–1935), branded it a "Bavarian Alpine Road" (Bayerische Alpenstraße). A neo-native, like many tourism advocates, Knorz had a day job as director of the Prien hospital. To further the idea of this road, Knorz launched a broad media campaign.[23]

Significantly, Knorz was not only a tireless promoter, but also an early member of the Nazi party. He joined the party in 1929, when it attracted relatively few voters. Knorz used the party newspaper as his platform to advertise the idea of an Alpine Road. He envisioned it as an educational tool to attain "quiet enjoyment of a beautiful, German landscape among Germans." At a time when more tourists took trips using automobiles, Knorz wanted to establish a home-based version with a nationalistic twist. In the same breath, he presented the new road as a counterpoint to attractions in Austria and Italy, enticing tourists to spend their time and money in Bavaria. Citing Mussolini's sponsorship of the autostrada, Knorz hoped for Hitler to become the guiding spirit of this highway project.[24] Knorz's vision was the extreme right-wing version of roadmindedness.

With Hitler still in the wings, the Great Depression became a rallying point for the Bavarian Alpine Road. In September of 1932, Knorz convened a meeting of tourism leaders from all over Southern Bavaria as well as members of the state parliament. Importantly, participants at the meeting envisioned that only a quarter to a third of the estimated three hundred miles (five hundred kilometers) of Alpine Road would have to be built from scratch; for the most part, the new road would consist of upgrading existing local connectors. Still, Knorz insisted that only a "real Alpine road," one at higher altitudes, would become an attraction, and that it should not leave Bavaria at all. Again, speakers stressed the competition from Austria, in particular, the project of a road to the Großglockner summit. According to a newspaper article, the four members of the state parliament present at the meeting—conservatives, Social Democrats, and National Socialists—signaled support (but not necessarily commitment). An incipient Alpine Road lobby, made up of local and Southern Bavarian tourism supporters, had become visible. In addition, Knorz was able to stir interest among newspapers. He was an inexhaustible, single-minded promoter whose affiliation with an extremist party gave him a particular platform. Knorz described his idea as necessary and generally popular. However, not even the regional tourism association for the Chiemgau region was convinced. As its leaders argued in private, it was simply too expensive to be built in the foreseeable future. Publicly, some members of one motoring club embraced the idea.[25]

The conservative state government in Munich, however, remained unimpressed. In January of 1933, just before Hitler's ascent to power in Berlin, Nazi representatives in the Bavarian state assembly put the road project on the agenda. In its budget committee, they requested that the state government adopt the Alpine Road and seek unemployment relief money from Berlin.[26] Historians argue that the general attitude of the Nazis in the state assembly before 1933 was largely obstructionist and propaganda-oriented—efforts to ban kosher butchering had been some of those most visible.[27] Given the generally chaotic political scene, the assembly did not pass a single law in the year before the advent of the dictatorship resulted in the shutdown of state parliaments. Embracing the Alpine Road was yet another sign of support for an unrealistic, expensive idea at a time of great economic and political turmoil.[28]

In addressing this idea in front of the parliamentary committee, the highest-ranking highway official in Bavaria, the civil engineer Josef Vilbig, dismissed it. He deemed it "out of the question" that Bavaria would pay for the road or use money from the Reich, given a long list of necessary road improvements elsewhere. Alois Hundhammer, a conservative politician who later became famous as cultural minister in postwar Bavaria, seconded that it was more important to focus on upgrading transport networks in other places than to build new ones.[29] The Conservatives put forward a motion of their own, which requested in more general terms that Alpine roads be upgraded for tourism and for work creation efforts.[30] This was more in line with Vilbig's vision.

The engineer had, in fact, published a comprehensive report on the conditions of Bavaria's roads in 1925, replete with a ten-year plan for modernizing existing roads, mostly by widening lanes and paving the roads. In stark contrast to the plans for the Alpine Road, the treatise was based on year-long traffic counts; heavily traveled roads indicated a need for upgrading.[31] At its heart, the question was whether roads should follow traffic or generate it. Proponents of the Alpine Road, by and large, sought to create traffic. They were motivated by economic concerns about tourist revenue and cultural arguments over a new experience of landscape. Like his counterparts in American state highway departments, however, Vilbig aimed to respond to existing needs and extrapolate from them—an engineering approach based on the assumed supremacy of numbers over politics. Notably, Vilbig had toured American roads after representing Bavaria at the International Road Congress in Washington, DC, in 1930. To be sure, Vilbig's report was not disinterested either: he described the growth of automotive transportation as a given. At a time when just one in 390 Bavarians owned an automobile, he predicted that his state would follow the path of other, wealthier regions and countries, as if it were a law of nature.[32] Vilbig's profession, his department, and civil engineers as a whole would (and did) benefit from such highway plans, and from motorization in general; such predictions were as much a precondition to motorization as a response to it.

When the proposal for the Alpine Road was on the floor of the Bavarian state assembly in early 1933, Vilbig compared costs and benefits, concluding that it was "impossible" for the state of Bavaria to become the main sponsor of what many considered to be a "luxury project."[33] For the Nazi faction, such expert reasoning was pusillanimous. Ludwig Siebert, the subsequent governor of Bavaria, thundered that it was the "damned duty" of the state to take leadership, especially when issues with thorny interests clashed.[34] Another Nazi delegate claimed that the goal of the motion was to force the state administration into showing whether or not it was willing to do "something generous" in economic matters.[35] A Communist state parliamentarian supported the project, even though, in his eyes, support for tourism only benefited capitalist interests. Mockingly, he wondered whether the Nazis wanted the Alpine Road as a better connection to their fellow Fascist Mussolini in Rome. Given the economic and political malaise of the late Weimar Republic, the grandiose Alpine Road was unrealistic and low on the list of priorities for a somewhat sober state government. At the same time, the Nazis embraced it as a slightly daring, modern scenic infrastructure with potential benefits for tourism and national reawakening.[36]

With the Nazi takeover of power in Berlin in late January of 1933, marginal projects such as the Alpine Road moved to the top of the agenda with ease. In April, the Bavarian state assembly met for the last time before its divestiture. Some of its Social Democratic members were released from prisons and the Dachau concentration camp for the occasion, only to witness a Nazi spectacle celebrating the end of democracy.[37] During his self-congratulatory speech, the new Nazi governor Siebert announced that the project of the Alpine Road would come to fruition and that it had been launched by a Nazi, namely Knorz.[38] In a long memorandum from the spring of that same year, Knorz indulged in references to the "racially still healthy" stock of rural Upper Bavarian residents. His main motivation for the Alpine Road was to convince tourists, both German and foreign, to spend more time in Bavaria by offering them an all-Bavarian road. Costs should matter less; above all stood the "effect on the traveler."[39]

Knorz's insistence on the landscape effects was tied to political arguments: he accused the last democratic Bavarian state government of having hid behind a stance of frugality and gave the Alpine Road project a Nazi lineage. Knorz was a member of the party, the idea had first been announced in a Nazi newspaper, and now the Führer was about to realize it. By late March, Hitler was quoted in local newspapers: the new road was not just a matter for Bavaria, but for the entire Reich, decreed the dictator. More importantly, it enjoyed his support.[40]

For activists such as Knorz, the Alpine Road was a political wedge and a token of his party's activism. But it does not follow that roadbuilding was necessarily a Nazi effort. When Frankfurt's interwar lord mayor, Ludwig Landmann (who was Jewish), championed the cause of the Hafraba interwar autobahn lobby, his support did not make the highways a Jewish issue either. Landmann, a member of the liberal German Democratic Party, saw the roads as a potential economic boon for his city.[41] Knorz, however, wanted to spur touristic growth along racial lines and further his party's cause at the same time. The Alpine Road's origin and support rested on National Socialism.

In a dictatorial gesture, Hitler made the Alpine Road one of his pet projects in early 1933. Local tourism leaders in Southern Bavaria still doubted
whether a southern route—higher up the mountains and further from established towns—was preferable.[42] Voices from Northern Bavaria preferred
a less expensive subalpine road in the valleys as late as the fall of 1933.[43] But
such concerns, based on sectionalism and economic concerns, mattered little in the new dictatorship. In a meeting with governor Siebert in September 1933, Hitler affirmed that he would give the project high priority; it had
to become "something really big, a road that the world would pay attention
to."[44] According to one newspaper, a "gigantic project" was in the works.[45]
Hitler's new regime was here to impress, and a Brobdingnagian road in the
mountains was one of the more visible signs of the dictatorial era.

Vilbig, the former critic of the project, now instructed district engineers
all over Southern Bavaria to supply detailed plans. He proved to be a loyal
civil servant. Like many officials, Vilbig joined the party in the spring of
1933. Interestingly, his design instructions all but did away with his former
reservations and concerns about costs. Not only should the road open up the
beauties of the Alps, but the road itself should have a beautiful and grandiose
effect through routing, massive structures, and integration into nature.[46]
Construction and maintenance should remain "within economic limits," but
this goal was clearly incompatible with the others.[47]

The Nazis, apparently, took the notion of uplift quite literally. An engineering study from the fall of 1933 mentions new parameters. The former
nucleus of the project was to be upgraded. Instead of planning the most
simple and direct connection from Füssen to Linderhof castle, the new goals
were to lead the road into higher altitudes and to make it "grandiose in its
own design."[48] For the entire route, the new plans included eight mountain
passes at more than 5,200 feet (1,600 meters) above sea level and one above
6,600 feet (2,000 meters). Earlier routings had included one pass at 4,700
feet (1,430 meters). In landscape terms, the higher elevations of the Alpine
Road were to create auto-touristic sights by traversing peaks that had been
the province of only transhumant livestock, herdswomen and -men, and
hikers.[49]

By late October, Hitler visited a giant three-dimensional model of the
Alps in the Bavarian Interior Ministry in Munich that had been made for him
and Fritz Todt, the top engineer of the Third Reich and the newly appointed
road czar of Germany.[50] Blue and red threads signaled two different routes,

Adolf Hitler and Fritz Todt studying a model of the German Alpine Road. Hitler and Todt were closely involved during the design phase of the German Alpine Road. Todt is pointing at routing alternatives. Viewing the mountain range from above, the dictator decreed that moving the road up into the mountains would serve its purpose of showcasing Alpine scenery. *Sonntag Morgen-Post. National-sozialistische Sonntagszeitung*, no. 44, October 29, 1933, 17–18

a shorter and a longer one.[51] Planners provided choices, but the last word was to be with the Führer, "like it is with everything that happens in the new Germany," as one newspaper put it. On its front page, a photograph of earnest men studying the mountainous landscape from above was entitled "The World Arms Itself—Adolf Hitler Creates Peaceful Achievements."[52]

Upon viewing the Alpine model, Hitler made the "opening up of all of landscape's beauties in the nature of the Alps" the top priority of all planning efforts. Providing jobs for the unemployed was mentioned, but the main rationale for the "great, creative achievement" was a scenic, automotive, tight embrace of the German part of the Alps.[53] It was not meant as a mere transportation route but as a means to produce and frame vistas for drivers. To achieve such scenic effects, Hitler desired even higher altitudes for the road and preferred the conquest—not the circumvention of—passes and peaks. For example, Hitler wanted to move the western portion of the road in the Allgäu region up to the Hochgrat summit, with an altitude of 6,017 feet (1,834 meters) above sea level. Elevation mattered, even if Austria and Switzerland would always best their German neighbor. Although not the highest Alpine road, it would be the longest and "in its variety the most beautiful" of its kind, the newspaper assured its readers.[54] A 1934 newspaper article described an Alpine Road that would traverse half a dozen peaks of more than 5,000 feet (1,500 meters); however, none of these new roads on mountain passes were ever built.[55] In general guidelines from the fall of 1935, Todt stipulated that the Alpine Road was to open up the mountains. While it surmounted valleys and peaks, it should never be allowed to run roughshod over nature. Nothing less than a masterpiece worthy of Hitler's name should be created.[56]

Construction on the Alpine Road began in November of 1933. Instead of the long-debated Füssen-Linderhof route, workers started blasting Alpine rocks in the very east of Bavaria, close to Berchtesgaden, and in the west, near Lindau. Although earlier promoters had thought the connection to Linderhof to be the kernel of the entire Alpine Road before the Nazi takeover, it was not built under Nazi rule or afterward.[57]

Rather than connecting to the Romantic dream castles of a nineteenth-century king, the Alpine Road's first completed stretches led to the mountain redoubt of a twentieth-century dictator and his henchmen. By the summer of 1936, motorists took to the new road in the vicinity of Berchtesgaden. Its most prominent user was Hitler. The two patrons of the Alpine Road, Hitler and Todt, had more than merely propagandistic interest in the project. The autobahn, from Munich to the Austrian border, provided a quick connection to Hitler's Alpine residence on the Obersalzberg, the site of the infamous summit with British prime minister Neville Chamberlain in 1938. The failed politics of appeasement are tied to the transformation of this mountaintop into a Nazi residential landscape with scenic surplus. Todt, the Reich's top engineer, vacationed regularly in neighboring Ramsau, where he bought a summer residence. Both locations benefited from the new road connections offered by the autobahn and the Alpine Road.

The Nazi infrastructures transformed the Obersalzberg, but it had not been a wilderness by any means. In fact, local innkeepers had developed a thriving tourism community with some prominent guests for several years. A frequent visitor from Vienna was Sigmund Freud, who wrote portions of

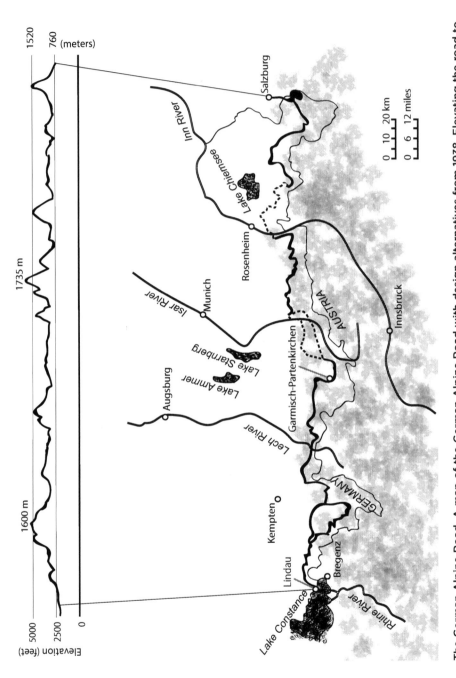

The German Alpine Road. A map of the German Alpine Road with design alternatives from 1938. Elevating the road to new heights was a result of Nazi planning. Most of the planned mountain passes remained unbuilt. Map created by Caitlin Burke, based on Michahelles, A.[ugust], "Die Deutsche Alpenstraße," *Zeitschrift des Vereins Deutscher Ingenieure* 82, no. 37 (September

Civilization and Its Discontents during his last stay there, a few years after
another guest had written parts of *Mein Kampf* on the mountain.[58] When
Hitler and his coterie of Nazi leaders monopolized the peak with its farms
and inns after 1933, locals were expropriated and the mountain became a
gated, elite gathering place. Branching off from the public part of the Alpine
Road, the connector to Hitler's residence (known as the "Eagle's Nest") was
a top priority, as Todt instructed the local engineers.[59] As Alwin Seifert put
it, Todt's goal was to offer his Führer an "outstandingly beautiful approach"
to the Obersalzberg.[60] In what amounted to constructional reverence, Todt
instructed local road builders to use "utmost diligence" when designing.[61]
After one of Hitler's architects presented plans for the house, Fritz Todt hiked
the mountain himself and came up with general plans for a winding mountain road. It terminated just below the summit, where an elevator took guests
up to the residence.[62] The roadbed was dynamited out of the mountain in
just over a year. Construction continued during the winter—unusual for a
mountain road—which made for particularly harsh working conditions.[63]
The expensive project, which employed up to 3,500 workers, served Hitler's
representational desires.

During the Nazi era, Hitler's mountain retreat close to Berchtesgaden
was covered extensively in newspapers, both domestically and abroad. The
huge window framing the Alpine panorama impressed Chamberlain as well
as millions of newspaper readers and moviegoers watching newsreels. The
propaganda machinery of the Third Reich took pains to portray the dictator
as a benevolent leader with a mountainous redoubt. Pictures of Hitler play-
ing with his German shepherd on the Obersalzberg were ubiquitous. The
most ardent among the Hitler admirers sought to get close to Hitler by visiting the mountain. Since it was off-limits to anyone not on official business
by 1937, even travelers struck with adulation for Hitler could not visit their
Führer there. Responding to popular demand and to the political realign-
ment of the Eastern Alps after the annexation of Austria, Nazi planners rearranged the eastern terminus for the Alpine Road.[64]

Incorporating some formerly Austrian territory, sixteen kilometers of a
new road overlooking the Obersalzberg were, in effect, a consolation prize
for those who could not visit Hitler's mountain retreat up close. Instead of
climbing some German peaks and ending at the fjord-like lake called Königssee, the Alpine Road would find its end point on a mountain overlooking

valleys and the Obersalzberg. Visitors could see the Obersalzberg that they could not visit. Such a trip would make it clear "why the Führer has chosen the Berchtesgaden area for rest and recreation."[65] Construction on the Roßfeld part of the Alpine Road, with an elevation of 5,200 feet (1,600 meters), began in the summer of 1938, with several Austrian engineers who had gained experience on the Großglockner road supervising more than three thousand construction workers and a contingent of fifty Italian stonemasons working on stone trimmings for bridges. All but one kilometer of the Roßfeld road were finished when Germany began World War II. Two inns offered views of the Eagle's Nest dwarfed by high Alpine mountains. With the German conquest of Europe starting in 1939, workers were drafted into the Wehrmacht and the road sat idle, except for an anti-aircraft unit stationed there during the last year of the war.[66] Local politicians convinced the Bonn government after World War II to close the gap and complete the road. The mountainous view from the road has served the movie industry, too: the blockbusters *The Sound of Music* and *Indiana Jones* both contain scenes filmed in the vicinity of the road.[67]

However, another scenic infrastructure remained unbuilt. As one historian explains, a proposed cable car on the Watzmann mountain, visible from the Obersalzberg, drew the ire of Alpinists and conservationists. Tourists on a Nazi pilgrimage to Hitler's mountain resort would have provided ample customers, but the opponents were able to prevent the structure from being built.[68] The Obersalzberg received Hitler's imprimatur in a physical and symbolic sense, and the Alpine Road was one of the features of this transformation of a remote tourism community into a site for diplomatic gatherings and popular adulation of dictator and scenery alike.

Celebrating the Alpine Road

When the first stretches of the Alpine Road were opened in 1936, the regime celebrated them as well as the larger idea of the road. Knorz, who had given the road a Nazi pedigree, had died the year before.[69] His ideas and public persistence before the Nazi takeover were overshadowed by Hitler's patronage, the new scale of the road, and changes in design. In a guidebook for the road—now called the German Alpine Road rather than the Bavarian Alpine Road—Todt conceded the initially slow pace of construction. However, this was because the first plans had been too "timid" in their approach to the mountains. Therefore, the first construction sites served as testing

"Happy Summer Weeks in Southern Bavaria" (1936). In this poster, southern Bavaria has become a technified landscape populated by a tourist couple in hiking gear. The farmhouse, the church steeple with its Baroque onion dome, and the maypole featuring a swastika flag dominate the foreground. The massif in the background is adorned with a cable car and a sinuous mountain road, both leading toward the peak. Bayerisches Hauptstaatsarchiv Munich, StK 6999

grounds to gain experience for building mountain roads. Based on these experiences, engineers could now aim higher and redesign the entire project so that they could "protrude into the heights much more boldly" than the old project.[70] Instead of building a road which nestled closely to the mountains, by 1938 civil engineers planned a road that was more domineering, less curvy, and more predictable. The plans included stretches as high as

5,700 feet (1,700 meters). In comparison, the average elevation on the Blue Ridge Parkway is 3,000 feet (900 meters), with the highest locations at 6,000 feet (1,800 meters). The Alpine plans called for 105 bridges, 15 tunnels, and 10 viaducts. For its Western section, a stretch of twenty-three miles (thirty-seven kilometers) at a minimum of 3,300 feet (1,000 meters) of altitude was a part of the design.[71] Costs skyrocketed, thus postponing the completion of the road. By 1939, the German Reich's priorities of war and European conquest meant that resources for roadbuilding were diverted to purely military purposes or those, like Hitler's mountain resort, deemed as such.

Todt also addressed the relationship between hikers and motorists in the Alps and tried to assuage conservationists. Hikers had been upset by the dust stirred up by cars, he stated; thus, providing a modern, dust-free road would nullify these concerns. Rather, the Alpine Road would create opportunities to hike with the automobile. This was not at all what purist hikers had in mind. As the historian Rudy Koshar observes, the writer Heinrich Hauser introduced auto hiking (*Autowandern*) as a motorized immersion into the entire range of sensory experiences offered by landscapes.[72] The idea of auto hiking was not a Nazi invention; Todt, however, gave it a particular Nazi twist by emphasizing Hitler's prioritizing of this infrastructure. As far as conservation was concerned, Todt stated that the road was simply too small to upset the grandiose Alps. Still, it had to be kept free of "fairground" architecture.

Automotive hiking required drivers to slow down. While the interstate highways of the postwar era enabled a steady flow of cars and trucks at high speeds (and the speed limit on the German autobahn was dropped in the 1950s), these landscaped roads were designed to decelerate and make motorists stop repeatedly. Hairpin curves had been a feature of mountain roads for more than a century, but viewing platforms and rest areas were specifically tourist-oriented accoutrements of such scenic highways. Through the design of the road and in publications, travelers were strongly encouraged, if not educated, to pause, rest, and admire. A Nazi guidebook for the German Alpine Road admonished its readers: "You can choose to push down the throttle and to compress all of this [the Alpine world] to a few hours, as if in fast-motion. But nothing keeps you from stopping the flow of pictures and to linger at those points which you deem the most beautiful."[73] The reference to cinema and moving images is obvious and not novel, as seen before. While they evoke consumer choice and the quasi-directing of one's own sce-

Auto hiking on the German Alpine Road. These two well-dressed female motorists appear to have disembarked from their vehicle at a hairpin bend of the German Alpine Road. By design, the highway slowed drivers and passengers down. Guidebooks admonished them to stop and combine driving with hiking in what some observers called "auto hiking." Hans Schmithals, *Die Deutsche Alpenstraße* (Berlin: Volk und Reich, 1936), image 135, page 108

nic road movie, these roads, however, prescribed views. The curves of the German Alpine Road and the relatively narrow roadbed made speeding most difficult. Stopping was only possible when rest areas allowed for it or traffic was sparse. Views from the road were normed, the result of design, planning, and construction. The trip was cinematic to the extent that it allowed for wide vistas, but the experience was as choreographed as a theatrical production.

In guidebooks, motorists were reminded that the landscapes they experienced were German and essentially so; their road trips were supposed to reaffirm their belonging to an ethnic collective whose cultural values were expressed in its landscapes. In the written equivalent of a wagging index finger, Todt instructed the drivers on the Alpine Road to be "quiet, considerate in conduct, and reverential toward the grandiose nature surrounding you." He also admonished them to thank Hitler. A breathless paean to the

A bridge spanning a valley on the German Alpine Road. Bicyclists and motorists appear to use the structure as a vantage point. Hans Fischer, *Bayern links und rechts der Alpenstraße* (Munich: Bergverlag Rudolf Rother, 1938), 49; photograph by Ernst Baumann

road in the foremost Nazi newspaper *Völkischer Beobachter* claimed that the workers wrested the road from the mountains to create an everlasting monument to the dictator.[74] Guidebooks placed baroque churches right next to valleys and mountains, thus creating a seamless web of nature, technology, and culture.[75] Drivers, however, had to meet racial criteria to qualify for rides. As noted earlier, Germans classified as Jewish were banned from driving by 1938.[76]

Controversies over Hitler's Project

The Alpine Road was clearly a dictatorial project. Its first stretches in Bavaria's southeastern corner served Hitler himself. His grandiose understanding of roadmindedness ruled. In addition to the dictator, however, less powerful drivers and passengers would use the road. As conflicts between planners and locals show, the intended users were tourists, not the residents of the areas connected by the new road. It is clear that Hitler and Todt were not driven by economic concerns, took little note of local complaints, and were eager to impose their centralist, metropolitan vision of a scenic Alpine road onto the landscape and its residents. While the rulers brooked no public dissent, a few instances of disagreements have survived in the archives.

For one, the Alpine Road's location on the ridges rather than in the valleys made it less useful for locals. Local farms would suffer as parts of their property would be taken, argued a petition signed by twenty-two residents of the Allgäu region. Business owners throughout the valley feared a loss of customers. Therefore, the locals pleaded with the authorities to upgrade the existing road in the vicinity of the town of Simmerberg rather than building a new one.[77] No answer is to be found in the archival files. (The new road was eventually built on its southern, more mountainous alignment.) Similarly, a motoring club was rebuffed when it suggested a more utilitarian alignment of the Alpine Road close to Lindau by connecting it to a local train station. This would not have been scenic enough, decreed the Bavarian state government.[78] A local resident's obvious, but apt, remark that a road on higher elevations would be unusable during the winter and thus offer less practical value went unanswered. Scenic Alpine roads were and are seasonal roads, at least in part.[79]

In the face of municipal requests, Todt defended his vision of a scenic infrastructure serving touristic visual gain, not local desires. Town leaders from Isny in the Allgäu region hoped for a connection to the Alpine Road, lest they be cut off from future tourist traffic. Todt conceded that their request was understandable. However, only the more mountainous routing would offer the long, sustained views of the Alps that the road was designed for. Occasional and short glimpses from the car would not suffice. The German Alpine Road, decreed Todt, was not a local connector, but a German high Alpine road competing with its ilk in Switzerland, Austria, and Italy. Therefore, local desires were less important than the "rules made by landscape itself."[80] No stranger to directives himself, Todt invoked a higher authority in announcing these decisions, which were as human as any.

Locals received the message of the new priorities loud and clear. Whereas requests for roads had to display at least a semblance of economic rationality in order to please Munich officials before 1933, dictatorial realities commanded allegiance to beauty, which—in the eyes of Todt and Hitler—was enhanced by elevation: the higher the elevation, the longer and wider the views motorists could enjoy. Most of the vista-friendly locations were to be created by building motor roads across mountaintops. In one location, an existing mountain pass built in the late nineteenth century, at the apex of a former medieval salt trading road, was renamed "Adolf-Hitler-Pass" and became part of the Alpine Road plans.[81]

During the Nazi period, conservationists were not involved decisively in the planning of the road. The Bavarian-wide office for conservation, a semi-official advisory body, did not challenge the idea of the Alpine Road as such, but suggested building a hiking trail from Lindau to Berchtesgaden in addition to the road, since the enjoyment of nature by hikers "is severely curtailed if the path has to be shared with the dashing automobile or the popping motorcycle."[82] Although organized hikers did not like the idea of the Alpine Road, they did not dare protest a project endorsed by the Führer himself.[83]

Todt was adamant in his efforts to control the roadside personally. Based on archival files, it appears that local authorities often considered the Alpine Road an opportunity for local businesses. Todt, on the other hand, wanted to ensure that the road would repel, rather than attract, roadside development. Usually not prone to polemics, he warned a local administrator that the road would become the "bane of the entire area" if construction were allowed to proceed unchecked and without the correct aesthetic convictions in previously undeveloped areas.[84] When locals close to the town of Inzell planned to use the new road as the site of a new inn, school, church, and apartment building, Todt decreed that these buildings could only be erected on another road, not on the tourist road.[85] Clearly, the Alpine Road was not to be a local road.

Even for the interventionist Todt, controlling every roadside building was not feasible—although the files contain several efforts by the highest-ranking engineer of the Reich to prevent large developments, such as vacation homes for the Siemens company or the dictatorship's tourist agency, *Kraft durch Freude*.[86] When the SS built a vacation home right next to the Alpine Road, Todt made sure that the power transmission lines would be buried underneath the road, so as not to mar the view of motorists and vacationing members of the terror squad. (It is now a youth hostel, with the same view.) [87] Todt and the SS did not bother with the regular local permit process and handled design questions on their own. Prisoners from the Dachau concentration camp—many of them Jehovah's Witnesses—were forced to build the access road for the hotel and were responsible for its maintenance. After the beginning of the war, it was used as a hospital for wounded SS men.[88] Elsewhere on the Alpine Road, presumably insufficient landscaping of a single car repair shop did not escape Todt's scrutiny.[89] While he showed no signs of tiring from such micromanagement, he did bemoan the lack of comprehensive, foresighted planning.[90]

In a 1937 assessment, a local conservationist subscribed to the general idea of the road as a provider of views for motorists. Therefore, it should be located not immediately adjacent to, but at some distance to landscape highlights in order to create vantage points. In one particular instance, the conservationist requested that a local lake was best enjoyed while viewing it from a distance. Road planners had situated the Alpine Road right next to the lakeside—this would be tantamount to a destruction, not an opening up of the landscape, according to the conservationist.[91] Such voices, however, remained marginal during the planning process.

Hiking and the Alpine Road

Despite the claims of peaceful coexistence between hikers and motorists, tensions remained. While the Nazi newspaper, the *Völkischer Beobachter*, had categorically declared that cars and motorbikes had replaced the placid hiker, since "only a few people" still had time to travel on foot, others were not as certain.[92] Against the advice of mountaineering organizations and local innkeepers, the state-wide tourism agency began to push for a high-altitude hiking trail connecting Bavaria's peaks in 1937. While such trails existed in Switzerland, Austria, and the Black Forest, Bavarian Alpinists disliked the idea because it would attract inexperienced hikers; local property owners declared the idea simply unsuitable for the Bavarian Alps. Tourism promoters, however, came up with a detailed plan to de-skill the hiking experience. All hikers would be able to use it without any danger: after each hour of hiking, a rain shelter would be available; every other hour, a shelter would provide emergency overnight accommodation; and every four hours, a mountain hut with a restaurant and bedrooms would greet tourists. Leisurely hikers could thus traverse three hundred miles (five hundred kilometers) in four easy weeks.[93] Such a predictable, almost undemanding hiking infrastructure would open up Bavaria's mountains to more hikers with less experience. In comparison to the more austere Appalachian Trail, this path would have offered plenty of food and rest options on the trail. While the tourism promoters mentioned that the Hitler Youth and *Kraft durch Freude* recommended hiking as a Nazi activity, the plans remained stuck in committees.[94] The suggested dynamiting and construction activity remained on paper. Almost half a century after the war, a named hiking trail began connecting Lindau and Berchtesgaden, but it amounts to a consecutive signage of existing trails—without the regularly appearing huts and shelters envisioned earlier. Rather than alluding to the Alpine Road, the Alpine Club remembered a trip taken by the Bavarian king Maximilian in the 1850s, whose approximate route this so-called "Maximiliansweg" has been following since 1991.[95]

The German Alpine Road in a European Context

The Alpine Road, like other scenic highways, was a response to other scenic infrastructures—and an effort to surpass them. Supporters of the project did not tire in pointing out that other Alpine countries had already built or were about to build scenic highways in the mountains. According to one observer, Italy had invested hardly any money in trains and cable cars but built roads in the Dolomites and around the subalpine Lake Garda.[96]

Although it would be difficult to know how many travelers used the Alpine Road, it is likely that its eastern sections were the most popular. One reason for increased Alpine visitation was not the attraction of the road itself, but the result of an administrative act. In May of 1933, the Nazi regime required any German traveling to Austria to pay a fee of 1,000 Reichsmark, which made it prohibitively expensive to visit. The goal was to hurt the Austrian economy, which depended on (German) tourism to a considerable degree. Predictably, tourist visitation to Austria decreased dramatically.[97]

The annexation of Austria in March of 1938 did away with national borders (the exorbitant fee had been dropped three years earlier). It also led to a mountainous mobilization, since the formerly Austrian Alps were now part of a pan-German domain. The highest peak of what the Nazis called "Greater Germany" (Großdeutschland) was no longer the Zugspitze in Bavaria, but the Großglockner peak on the border of Tyrol and Carinthia. As a monument to Austria and its shrunken post–World War I Alpine territoriality, the Republic of Austria had begun to decorate the Großglockner with a High Alpine Road; it was opened in 1935, as discussed in chapter one. Hugh Merrick, the widely read author of a book on Alpine highways, recognized and reinforced the idea of the road as a national monument on par with the Swiss Jungfraujoch mountain saddle, accessible by rail since 1912. He praised it as "magnificently daring in conception, superb in execution, and positively staggering in its furnishings." But he also averred that the new Großglockner road was marred by "an element of superb but conscious showmanship."[98] In the end, for Merrick, it amounted to "a glorious piece of window-dressing." He abhorred the masses partaking of the Alpine scenery in buses.[99] For Alwin Seifert, the main problem with the Großglockner road was the visual dominance of restaurants, huts, and snack stations which "devalued" the landscape; polemical as always, he compared it to a gold rush town.[100] The road did attract many tourists, including those who did not own cars. Almost one out of ten visitors on the Großglockner Alpine Road in summer of 1939 traveled with *Kraft durch Freude*, presumably as passengers in the hundreds of buses climbing the road.[101]

The most basic reason for building the German Alpine Road—avoiding Austria and channeling tourist revenue back to Germany—became moot with the annexation of Austria. But this would not stop the planners in Berlin from pursuing the project. Rather, they sought to incorporate even higher mountain peaks in former Austria into their designs. For example, Bavaria's westernmost city, Lindau, was to lose its connection to the German Alpine Road. Instead, the road would now lead over the Austrian Pfänder peak (3,490 feet, or 1,064 meters), right outside of the city of Bregenz and a vantage point for overlooking Lake Constance. Lindau's lord mayor protested in vain as Todt recognized a "unique opportunity."[102] Territorial gain, in this case, meant elevation gain, and Todt was eager to exploit it. With the beginning of World War II, all of these plans came to naught.

While the Alpine Road transformed parts of Southern Bavaria for touristic goals, it is worth noting that Nazi Germany used its most extensive and violent landscaping visions for Central and Eastern Europe. After the conquest of Poland, landscape architects, geographers, conservationists, and other planners produced detailed plans in the early 1940s to remake the environment of parts of Poland and the Soviet Union for German settlement. These imperialist projects were predicated on murdering Jews living in the area and subjugating Slavs. Murder and environmental transformation would go hand in hand with the goal of creating productive agricultural landscapes. Directly referencing Western expansion in the United States, Hitler declared that "the Volga must be our Mississippi."[103] These brutal plans were secret (and never carried out), but they show how the public remaking of landscapes in the case of the Alpine Road was only one outlet for the transformationist visions of Nazi Germany. The road in the mountains was to celebrate existing landscapes; the plans for the "Eastern territories," however, rested on obliterating the landscapes and on mass murder to remake the area for German colonists.

Transatlantic Connections

While public proclamations stressed the German character of the Alpine Road, engineers continued to pay close attention to roads and landscaping in the United States. In addition to the usual exchanges via journal articles and conference visits, Todt hosted Arthur Casagrande, a Harvard specialist in soil mechanics, for a "brief consulting" visit in the spring of 1934.[104] The professor's brother Leo was working for Todt's agency on soil mechanics as well.[105] (Soil mechanics refers to the study of the dynamic underbody of roads.) The pace of visits to the United States by German civil engineers continued unabated.[106] The reverse was true as well: both in 1936 and 1938, Bureau of Public Roads Chief Thomas MacDonald toured the autobahn and other German roads. Todt hosted him personally and provided a tour guide.[107] While impressed by the scale of German roadbuilding, American engineers pointed out the mismatch between the low number of cars and the mileage of new roads. "Germany has the roads while we have the traffic," as Michigan's highway commissioner noted poignantly.[108] A paper by Gilmore Clarke on Westchester County bridges can still be found in the files of the Nazi road administration, as are publications on the 1939 World's Fair in New York.[109]

Small as it was, the Alpine Road still served as a showpiece. In 1934, the International Road Congress took place in Munich. Delegates from all over the roadbuilding world exchanged their views as they had done before in other locales; the Nazi regime was eager to present its plans and construction sites for the autobahn and the Alpine Road. In his welcoming address, Hitler's deputy, Rudolf Hess, presented the two roads as tokens of energetic Nazi planning originating with Hitler, not as apolitical infrastructures. In turn, Todt put these roads in the context of Roman and Inca roads, thus imbuing them with an aura of imperialism. At the conference receptions, delegates indulged in 1,790 quarts (1,694 liters) of beer and 2,252 cigars.[110] When the congress went on tour, seventeen hundred participants were bused all over Southern Bavaria in a single day—the organizers had to requisition buses far and wide to accommodate the crowds. Some delegates observed the new roads from a zeppelin airship.[111] The construction sites for the Alpine Road and the autobahn were showpieces for the international road experts akin to the presentation of the Mount Vernon Memorial Highway at the Washington congress four years earlier.

When New Deal promoters produced a brochure on "Roadside Improvement" summing up their experiences and recommendations, Todt's office was quick to get hold of it and have it translated. To his great chagrin, however, he realized that engineers in the field sometimes had not received copies of the translated American report.[112] In the summer of 1936, Todt had to conclude that the admonishments and presentations on landscaping, whether American or German, had not been heeded at all and had no practical effect on mountain roads.[113] An engineer responsible for the western part of the Alpine Road spoke of the failure of instilling landscaping ideas in younger civil engineers and contractors. Given their predilections and training, these engineers were overwhelmed by the architectural language used in such publications and struggled to translate the advice into designs. Pictures of good and bad examples without long explanations would be much more effective, he suggested.[114] But no German version of Jac Gubbels emerged.

When all else failed, Todt's heavy hand intervened. Ever the expansionist, the Third Reich's chief engineer wanted to adorn German landscapes beyond the Alps with more scenic highways. For Todt, roadmindedness was not limited to the Alps. Over several years, he pushed for extending an existing high-altitude tourist road in the Black Forest in his native state of Baden. Local conservationists tried to delay the project by declaring one of the summits a nature preserve. But after 1939, portions were built as a military road, given the region's proximity to France. Tellingly, in one of the many letters exchanged with regional offices regarding the road, Todt cited American parkways as examples to emulate. Given that Hitler was to provide the German masses with affordable Volkswagens, motorists should not be excluded from the most scenic areas, Todt opined. Conservationists and hikers' organizations could not disagree more. To no avail: Todt pointed out that "the most wonderful nature parks in America are generally traveled through only by automobile, and they are more beautiful and more lavish than our petty efforts to conserve some old tree or a tiny, limited area." Invoking the spatial largesse of the United States, Todt claimed that mobility-based conservation under his leadership would be preferable to privileging hikers searching for solitude.[115]

In his closed-door dealings with administrators and conservationists, Todt had to uphold his visions against the last vestiges of established bureaucracies and of civil society. Under these circumstances, the engineer held up the United States as a beacon of car-based scenery. His more general publications on the nature of technology and its role in the Third Reich, however, stressed a German-centric understanding of technology as a constructive force under Nazi auspices. While Todt was fluent in matters of American road-building and parkways, his public presentations emphasized the vernacular and the national. Apparently, the chief engineer of the Nazis never visited the United States himself. According to his biographer, he had been tempted to emigrate as a young engineer, fascinated as he was by the writings of F. W. Taylor and Henry Ford, and the prospect of putting a civil engineering degree to good use in a road-friendly country.[116]

American parkways had impressed Europeans with their wide right-of-way, which ensured control of the roadside. But even for the Alpine Road, the right-of-way was only as wide as the road and a few additional centimeters of roadside, legally speaking. Roadside control was about controlling the view for drivers and passengers. Without the American practice of purchasing the adjacent areas, some planners sought legal recourse. For Alwin Seifert, the solution was to elevate the road and roadside to the legal status of a protected area as outlined by the Nazi nature conservation law. Thus, roadside construction would be banned once the road was in place. Plans for hotels, inns, and souvenir shops would be nipped in the bud. The Nazi dictatorship had passed a fairly comprehensive conservation law in 1935, which conservationists used to protect natural monuments such as lakes and heaths, for example.[117] If the roads and roadside did not enjoy such status, "the most severe devastation of nature" would ensue, argued the ever-alarmist Seifert. Without protection, in a few years no one would understand why the road had been built: "Our grandchildren want their mountains back from us!"[118]

Seifert found a backer in Todt. He engaged in an extensive administrative battle with Hermann Göring, the second most powerful person in the Reich—whose purview included conservation. Todt pressed for the Alpine Road and its surroundings to receive conservation status. Lacking the legal authority to do so himself, he all but pounded the table and threatened to go alone. Göring and Todt were locking horns, while construction continued.[119] After almost two years of wrangling, they reached an agreement in the spring of 1938. But it did not really matter. The erratic pace of construction and Todt's interventionist managerial style—an American newspaper aptly dubbed him the "one man boss"—negated any legal procedures for the comprehensive planning that he envisioned.[120] Overall economic priorities shifted from war preparation to war by the fall of 1939. The Alpine Road remained too enmeshed in the personality-driven politics of the Third Reich, in particular Todt's and Hitler's, to serve as an agent for regional planning. A short attention span does not make good policy.

Planning the Blue Ridge Parkway

In comparison, the planning frenzy of the New Deal in the United States was both more varied and contested. A nationwide network of interstate highways was not part of these plans. While traffic intensified in urban corridors, the most extensive new highway construction project of the 1930s was the Blue Ridge Parkway in the Appalachian Mountains. It was to be a national road, to be sure, but one that gave access to and exhibited a selective vision of rural culture, mountainous landscapes, and largely pre-industrial ensembles of humans, technology, and nature. While the federal government and its metropolitan planners were clearly at the helm, state and local interventions shaped the project's technological landscape to some extent.

In obvious contrast to the Berlin dictatorship that tolerated no dissent, the politics of landscaping bore traces of debate and disagreement, if only behind the impressive facade of a mountain road.

While Bavarian leaders fought over new roads, tourism promoters in the American South sought to get their share of the parkway frenzy in the United States. The Appalachian highlands, remote and sparsely settled, solidified their status as middle-class vacation spots in the interwar period. In Virginia, North Carolina, and Tennessee, the counties around the spine of the Appalachian Mountains were among the poorer ones. Agriculture and silviculture predominated; pockets of industry existed but did not lead the economy. In the late nineteenth century, white Southern elites began to summer in the mountains, especially at resorts with springs. Higher-altitude scenery and the cooler mountain air enticed visitors. The relative gain in elevation and the multiple chains of mountain ranges in view were much celebrated.[121]

As was the case with other tourist destinations, two developments in the second half of the nineteenth century anchored tourism more deeply in the local economy: railroads and cultural stereotypes. In the case of Appalachia, the remote, rugged land met its match in the cliché of the mountaineer detached from civilization. A journey into western North Carolina could become a journey back in time. Tourism boosters spread such stories about the primitive locals and devoted equal energy to providing touristic infrastructures. By 1930, tourism had become the most important economic activity in the western parts of North Carolina, with Asheville as its center.[122] Hotels provided comforts for travelers, and Asheville's proximity to the highlands allowed tourists to hike or to partake of nature's wonders in less arduous fashion.

The chain of the Appalachian Mountains was only a few dozen miles from the amenities of Asheville. The top attraction was Mt. Mitchell, at 6,684 feet (2,037 meters) the highest peak east of the Rocky Mountains, as local tourism boosters did not tire to point out. A railroad, initially intended for logging, went up the mountain as early as 1911, but its operators realized a few years later that they could profit from hauling tourists as well. By the early 1920s, the first motor road went almost to the mountaintop and attracted exactly the kind of tourists that boosters wanted: white, middle-class, mostly urban visitors willing to spend money to ascend the peak without a major physical effort.[123] Remote as it appeared to outsiders, the Appalachian landscape was transformed for and by tourism, at least partially. Historians of the region maintain that logging had changed the face of the land most drastically up until the 1920s; afterward, tourism assumed this role.[124] Calls for "improving" southern roads were driven as much by utilitarian transport needs—getting farmers and their goods to (mostly urban) markets—and by desires to create touristic implements and scenic driving opportunities.[125] By the 1920s, Appalachia's visitation industry was eager to make the switch to a car-and-road version of tourism. New or improved roads were to bring more tourists to their hotels and to sights. Roads in Appalachia tended to be poorly maintained and were local connectors rather than long-distance routes. As early as 1909, a road convention in Asheville demanded tourist-oriented roads, some of which would be located on the spine of the Appalachians. One proposal was for a "Crest of the Blue Ridge Highway."[126] Since more people lived in the valleys and roads tended to connect larger settlements, such proposals remained farfetched until the Great Depression.

Local elites and tourism promoters understood that their region was competing with others over urban tourists with disposable incomes from coastal Carolina locations and cities such as Richmond and Baltimore. While their brochures extolled the sylvan wonders of nature, they knew that one important aspect in this competition was infrastructure, not just in terms of accessibility, but in terms of making the infrastructure itself scenic. The scenic policies of the National Park Service and the various parkways built around the country contributed to this Appalachian roadmindedness.

When the New Deal raised the prospect of federal funding, a bold proposal for a scenic road of several hundred miles landed on the desk of Harold Ickes, Secretary of the Interior and director of the Public Works Administration (PWA). State and federal politicians from North Carolina and Virginia put forth the idea as a way to fight unemployment and stimulate tourism. They also pointed to the growing momentum for a national park in Tennessee and North Carolina, eventually known as Great Smoky Mountains National Park. The new route would connect this new national park with Shenandoah National Park in Virginia, the establishment of which began in 1925. Because the federal government did not provide money for the latter, the state of Virginia began buying up land and displacing residents. Within Shenandoah, construction of the Skyline Drive had begun under President Hoover as a relief project in 1931. This 105-mile (169-kilometer) crest highway had been suggested by locals earlier and became subsumed into the work

Skyline Drive (Virginia), 1937. The Skyline Drive in Virginia was the first rural parkway built by the federal government. Thousands of locals were displaced for the construction of the road. This image shows Hezekiah Lam and his daughter (described as "hollow folk") waiting at a scenic stop for motorists to give them alms. Photograph by John Vachon, Library of Congress, Prints and Photographs Collection LC-USF33-001022-M2

relief efforts of the New Deal once Roosevelt took office. The as-yet-unnamed longer scenic drive would then extend Skyline Drive all the way to Great Smoky National Park in the form of a thin ribbon.[127]

Given the putative connection between two national parks, the proponents branded the road as a federal project worthy of federal funding. After some political wrangling, they succeeded by November 1933. Ickes's boss, Franklin D. Roosevelt, had embraced parkways when he was governor of New York (and even before), and saw their popular appeal and political purpose as a way to fight unemployment. Establishing national parks, even as a ribbon road, in the Eastern part of the country brought these recreational landscapes closer to urban centers and closer to becoming a reality.[128]

The Blue Ridge Parkway and other planned national parkways were not simply longer versions or extensions of the earlier parkways in the Northeast and Midwest. They were designed for higher speeds; traversed several different ecosystems, landscapes, and forest types; and varied in altitude. In considerable contrast to urban and suburban parkways, their intended users were not drivers living in the vicinity of the roads, but mostly metropolitan tourists who would venture out into the countryside for short jaunts or longer vacations. [129] As historian Timothy Davis points out, the national parkways of the New Deal exposed landscapes rather than hiding them, as some of the earlier suburban parkways had done.[130] The management of these roads by the National Park Service was not simply a bureaucratic matter accompanied by a change in design. This federal agency commanded considerable financial, political, and administrative support during the New Deal. Even when it waned during and after the war, institutional momentum ensured that the planning, building, and maintenance of these roads would continue. In other words, the Blue Ridge Parkway demonstrates the growing imprint of the nation-state on a landscape defined as remote.

The Politics of Altitude: Routing Disputes

Conceived by some local elites, financially supported by the federal government, and planned by civil engineers and landscape architects working for the National Park Service, the Blue Ridge Parkway became truly a national project. Before surveyors went into the field to stake out the new route, however, a political conflict shaped the general routing. As in the case of the German Alpine Road, elevation was at the heart of the matter: At what altitude should the new scenic road be built? The German dictatorship solved such queries with dictatorial simplicity and by fiat. In the American case, a more deliberative and public conflict took place. It had important implications for landscapes and driving.

The basic idea of connecting Shenandoah and Great Smoky Mountains National Parks did not imply searching for the shortest route and an alignment that would have been easiest to build. Rather, the production of scenery instead of ease of transport was to be one of its main effects. Since the Appalachian Mountains straddled Virginia as well as North Carolina and Tennessee in this corridor, the latter two states sought to bring the southern end of the road onto their territories. They competed for federal money and an infrastructure that would bring in tourists. Meetings with and letters to politicians, newspaper articles, letters to the editor, and finally a public hearing were the main arenas of exchange. In the process, the scenic qualities of the planned parkway were contested and negotiated.[131]

Several voices made themselves heard in this altercation. The most recognizable were delegations from the states of Tennessee and North Carolina. Tennessee's proposal was for a parkway reaching some peaks and bottoming out in some valleys, thus maintaining a variety of views to be seen from the parkway. It would terminate in Gatlinburg. In contrast, the routing preferred by North Carolina proponents placed much greater emphasis on ridge locations, which would allow for more and farther-reaching views. Their suggested route would end in Asheville. Apart from these sectionalist scenic preferences, landscape architects within and outside of the National Park Service, conservationists, hiking clubs, and public intellectuals took note and presented their suggestions. This was a national debate; scenery and driving were at stake. The well-publicized plans of the federal government for extensive federally funded parkways in the Appalachians set in motion an exchange over scenic roads, their proper extent and location, and their desirability. After the proliferation of urban and suburban parkways, their large-scale adoption by federal planners engendered criticism and calls for variety.

At the far end of the spectrum, one hiker decried "a good deal of a white elephant" and opined that "every man ought to have a job before scenic highways are constructed," when New Deal policy included fighting unemployment through road construction. Road boosters, unsurprisingly, competed with each other in praising automotive scenery constructed by roads. Between these two poles of road denial and road embrace emerged various voices. Among the most interesting was Benton MacKaye (1879–1975), the regional planner and wilderness advocate whose most widely known legacy is that of having started the movement for the Appalachian Trail, a hiking path mostly along the crest of these mountains running from Maine to Georgia, over some 2,200 miles (3,500 kilometers). The environmental historian Paul Sutter calls him "one of the most important and imaginative thinkers of the early twentieth century" in the United States.[132]

MacKaye's enthusiasm for hiking as a means of building new communities did not mean opposition to automobiles—far from it. He realized that automobiles could enable hikers to reach more remote locations than those served by railroads. But the growth of automotive traffic in the 1920s made him warier of what he called "gasoline locomotives." Rather than hoping that automobiles would deliver the country from the evils of industrialized transportation on the railroads, MacKaye feared that they would become just as destructive unless they were tamed. As a regional planner, he envisioned highways separated from urban centers, with control of the roadside and amenities for riders. A paper that he coauthored with Lewis Mumford envisioned the purgative power of cars and roads, if properly controlled and designed. Their 1931 *Harper's Magazine* article "The Townless Highway" suggested a network of parkways connecting major urban centers. The "divorce of residence and transport" was the main goal of these proposed new types of roads.[133]

Roads with wide rights-of-way and without unsightly accoutrements (the hot dog stand was cited again) would bypass cities and towns rather than cut right through them. The purchase of land and its regulation would ensure high design standards. New planned communities, garden cities for the automotive age, would help to disperse dense urban environments. Finally, such roads would be safer, too, since grade crossings would be a thing of the past. To achieve these goals, the federal government would award its subsidies to the states under the 1916 Federal Aid Road Act only if new roads were to meet these requirements. Thus, the "helpless and bewildered efforts of the past" would end. Such roads, according to MacKaye, were not the opposite, but the "complement in a sense" of the Appalachian Trail.[134]

On a larger level, MacKaye and Mumford believed in the early 1930s that cars and roads still had the potential to positively transform society and economy. Mumford had already praised Clarke's parkways in the New York environs for their aesthetic standards. In conjunction with the switch to electricity as a new way of distributing power, automobiles and roads, if properly managed, could help to usher in the "neotechnic" phase that Mumford envisioned. Cleaner, greener transportation technologies could supersede the dirty, industrial system of movement embodied by coal-burning railroads, which were part of the "paleotechnic" phase of "carboniferous capitalism" preceding the neotechnic in Mumford's sequence of historical periods. Scholars have correctly identified the technological optimism and circular logic of Mumford's thinking in the 1930s. Still, it is remarkable to realize that he thought the automobile's powers had not been fully used, mostly because automotive infrastructures had been grafted onto older ones: "All the mistakes that had been made in the railroad-building period were made again with this new type of locomotive [the automobile]," with the exception of parkways. Such nuanced criticisms went hand in hand with the hope that "the special habitat of neotechnic civilization" would be the uplands, whose healthier environments were within easier reach because of the automobile.[135]

Both Mumford and MacKaye had definite ideas on the design of roads. When it came to specific locations, their stance was more reactive than assertive. In particular, MacKaye was adamant that the ridgetops of the Appalachians should be reserved for the Appalachian Trail to let hikers enjoy solitude and scenery. Therefore, Hoover's Skyline Drive in Virginia drew MacKaye's wrath, since a large part of it sat right on the ridgeline. His protests against this particular road went nowhere. In response, MacKaye proposed an Appalachian Intermountain Motorway encircling the area from the Adirondacks to the Great Smokies—a much longer road than the one eventually built. This motorway was to remain in the valleys for the most part, reach up to some mountain sides, and only rarely cross mountaintops. As MacKaye's biographer reports, his efforts in convincing the Park Service to sponsor this road came to naught. In the early 1930s, MacKaye, one of the leading voices of the wilderness movement, wanted more roads, not fewer, as a response to both utilitarian highways and skyline scenic roads. For him, cars and roads needed management, not negation.[136] MacKaye and Mumford presented their own version of roadmindedness.

The publicized plans for the Blue Ridge Parkway threw such issues into even starker relief. At stake were several hundred miles of Appalachian ridges. When he learned of North Carolina's plans to let the parkway occupy the mountaintops, MacKaye crystallized his thoughts by distinguishing between "skyline" and "flankline" mountain roads. The latter would be just like his proposed Appalachian Intermountain Motorway. Their attraction was variety, as the road occupied different altitudes, thus allowing views both of and from the scenery. Flankline roads contrasted with skyline roads, which tended toward "monotony of view"; drivers' views would be directed away from the range rather than to it. Even worse, skyline roads would destroy wilderness while flankline roads would leave it intact.[137]

The historical moment for establishing a wilderness movement in the United States was planning for New Deal scenic roads, according to Sutter. A group of activists sought a new level of protection for areas without the visible imprint of humans. And this meant leaving them free from roads, which brought noise and exhaust fumes. Tellingly, the foundational meeting for the Wilderness Society took place in October 1934 at a roadside in Tennessee. Cars and wilderness were intertwined. As Sutter points out, "Wilderness preservation would have made little sense prior to the proliferation of automobiles, not only because the essence of wilderness was its resistance to mechanized transport but also because mechanized transport was itself essential to wilderness access." This tension was obvious to reformers like MacKaye, who imagined flankline roads as a conduit to regulate cars and roads.[138]

As a response to, critique of, and elaboration of parkway development, wilderness advocacy mattered, and that advocacy entered the fray over the routing of the road. Harold Ickes was the federal official with the power to choose a route favoring either North Carolina or Tennessee. He charged a committee to provide an expert report; it was chaired by one of his regional advisers and staffed by Thomas MacDonald of the Bureau of Public Roads and National Park Service director Arno Cammerer. In the political sphere, public pressure began to mount. At all these occasions, altitude was at the heart of the issue. The committee held two public hearings in 1934. Anne Whisnant, the foremost historian of the Blue Ridge Parkway, aptly concludes that these decisions over design were deeply political.[139]

As North Carolina's lobbying group pointed out, locating the bulk of the road in its state would allow for a mountaintop road similar to the Skyline Drive in Virginia. The state's chief highway engineer, R. Getty Browning, asserted that most of the route could be located above 5,000 feet (1500 meters), thus providing "the greatest amount of scenery." Staying on the ridgetops would require less cut and fill, Browning argued, thus decreasing the cost. In more florid language, a North Carolina congressman enlisted even higher powers: "Nature has fixed where this road should be located. . . . You must take the road to the scenery, you cannot take the scenery to the road." Referencing Europe's most touristic mountain range, the boosters claimed that the "scenery in the Grandfather [Mountain, near Linville, North Carolina] area is Alpine in its wildness and beauty."[140]

Tennessee's scenic point, however, was about variety. Its delegation suggested a combination of the flankline and valley roads with occasional visits to mountaintops. The lower stretches would provide opportunities for camping and offer "relief" from traveling on the ridges. The state's representatives also appealed to a sense of fairness and suggested dividing up the southern part of the parkway between North Carolina and Tennessee.[141] The latter's presentation prevailed, and the committee charged by Ickes recommended a routing split between the two states. After going into North Carolina for several miles, the road would move westward to the Unaka Mountain Range in Tennessee and continue southward to Gatlinburg, with a possible fork for both that city and Cherokee, North Carolina, outside Great Smoky Mountains National Park. The topography for this route would be more varied, the road alignment easier, and the scenery excellent. MacDonald, the road builder, and Cammerer, the Park Service director, were in agreement.[142]

A young landscape architect, Stanley Abbott, suggested the same route in the interests of variety and economy. He had been trained by and worked with Gilmore Clarke. The National Park Service hired him to plan the Appalachian road. Thomas Vint, the Park Service's chief architect, described the best possible route as "essentially a mountain route utilizing ridges, mountain slopes and mountain valleys. It is not an actual skyline drive."[143] Abbott admitted that a ridge drive would "offer unusual views of great power and beauty" but raised the "possibility that the tourist would become tired with 500 miles of mountain scenery." Later, he likened a series of panoramas as a "fortissimo," and therefore not as interesting as a "fortissimo mixed with a little pianissimo."[144]

Abbott's training as a landscape architect made him emphasize variety over monotony; drivers and passengers would be exposed to changing landscape features rather than mostly the view from above. Countering Browning's assertions, the architect averred that ridgetop locations would be more expensive to build, given the gaps between ridges, and costlier to maintain. Echoing MacKaye's concerns, Abbott noted a "lessening of the present recreational value of the wilderness areas" for a "considerable number" of people. In line with contemporaneous debates pushed by Mumford and others, Abbott's report went even further and suggested studying a valley location that would function not only as a seasonal tourist road, but as a year-round "passenger parkway" for locals as part of a "comprehensive regional plan." Such a combined road would have retained some of the design features of the Northeastern parkways, on which Abbott had been trained, and the Mount Vernon Memorial Parkway. As part of regional planning, it would take on the function of the townless highway that Mumford and MacKaye had proposed: a road that would contribute to refashioning settlement patterns and building new automotive-oriented garden cities. In other words, Abbott's report was fully immersed in the debates of the day regarding wilderness, landscape, roads, and planning.[145] Hiking groups echoed these voices, arguing that a lower-lying road would possess "superior utility and beauty."[146]

Given all these voices and proposals, the ultimate decision lay with Harold Ickes. The nascent wilderness lobby was delighted when Ickes requested

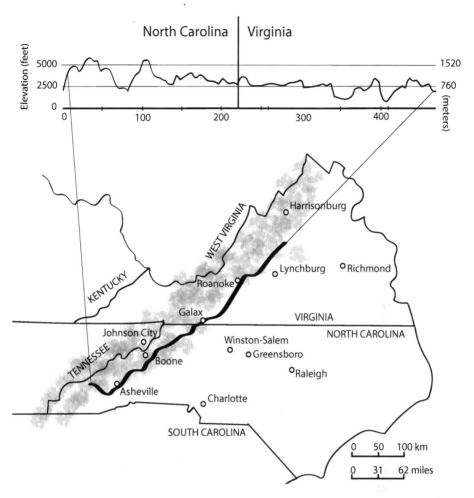

The Blue Ridge Parkway. This map shows the Blue Ridge Parkway as built. Both North Carolina and Tennessee fought over the southern terminus for the road, with North Carolina being victorious. Map created by Caitlin Burke

Robert Marshall, a forester in Ickes's Interior Department and later a founding member of the Wilderness Society, to provide him with opinions on road development in the national parks. In the fall of 1934, Marshall toured the Southeast and compared the two routes under discussion. After weighing the advantages and disadvantages of the Tennessee and North Carolina proposals, Marshall concluded that more important than taking sides on this political battle was the "necessity of keeping the parkway out of the few important primitive areas which are still left in this region" and listed three areas, in particular, to be avoided. These areas needed wilderness protection.[147]

No matter—the suggestions by Tennessee officials and the expert reports of MacKaye, MacDonald, Cammerer, Abbott, and Marshall went unheeded. North Carolina won the day. As Whisnant describes it, backdoor machinations and political dealing swayed Ickes and President Roosevelt more than professional testimonies. Namely, the expertise personified by North Carolina's expansionist state highway department and its chief engineer Browning trumped those of activists such as MacKaye and of the landscape architects in the employ of the National Park Service. The Interior Department conceded that both routes appeared equal "from a scenic standpoint." However, the main entrance to Great Smoky Mountains National Park would be Gatlinburg, and Tennessee was about to receive New Deal dollars through the Tennessee Valley Authority, which made it fair to award the southern end of the parkway to Asheville's tourism industry, according to this view. Lobbying from the established tourist interests in Asheville was more effective than comparable Tennessee efforts. In the case of the Blue Ridge Parkway, Marshall's request for establishing wilderness areas took a back seat.[148]

While the political process led to North Carolina's victory, the "continually unfolding panorama of magnificence" envisioned by that state's governor did not materialize quite that way. The Blue Ridge Parkway did not become a pure crestline road. For example, the various ridges of the Appalachians dip at certain points. Landscape architects and civil engineers on the ground took Ickes's instructions to follow the crest to be a general guideline. When it came to locating the road on the ground, factors such as topography, estimated cost of construction, availability of land, and negotiations with owners mattered. "We and the engineers together just drilled and drilled, all of us, on the business of following a mountain stream for a while, then climbing up on the slope of a hill pasture, then dipping down into the open bottom lands and back into the woodlands," recalled Abbott. The key words for him were variety through alternately concealing and revealing landscape features. About one-third of the eventual route was located in national forests, which allowed for higher elevations at no cost to the Park Service. When choosing metaphors to describe his work, Abbott spoke of painting, photography, and sculpture. In yet another reference to cinema, he compared his work to that

The Blue Ridge Parkway under construction. The road and its surrounding scenery were constructed, but they were also meant to blend into each other. The tunnel portal under construction showcases the meeting points between human work and geological forces. National Archives, College Park, Maryland, Special Media Archives, Still Picture Unit, 55–1813

of the "cameraman, who shoots his subjects from many angles to heighten the drama of his film."[149]

Still, dozens of miles of uninterrupted ridgetop location were the result of the political disputes. As government employees, landscape architects could deviate on details, but not on general policy. Consequently, drivers' and passengers' views from the Blue Ridge Parkway are often directed toward valleys below them and ridges in the distance. As the landscape writer Alexander Wilson puts it, "motorists feel that they are on top of the world," together with their car, and "in total harmony with nature."[150] Going southward from Shenandoah National Park, the road follows the crest of the Blue Ridge for sixty miles (97 kilometers). It then drops to its lowest elevation, 649 feet (198 meters), to traverse the James River. The parkway climbs up again, drops, and rises gradually at river valleys and other features. Its highest point, at more than 6,000 feet (1,800 meters), is located near the peak of Balsam Mountain in North Carolina. With precision, the Blue Ridge Parkway provides automotive access to all of the areas that Marshall had asked to be left untouched by the highway. Needless to say, portions of the road are seasonal and occasionally closed because of snow and ice.[151]

This is not to say that all of Marshall's interventions went unheard. A planned ridgetop road for the Great Smoky Mountains National Park never went beyond initial plans. The wilderness advocate sent several successful missives to Ickes, arguing against skyline drives. Instead, Marshall concurred with Park Service plans to build a road to Cades Cove, a broad valley flanked by mountains showing signs of logging and agriculture, and "one of the least wild sections of the park, and yet it has immensely impressive scenery." The idea was to provide an outlet for the pressures built up by automotive tourism and its lobbies, and to reserve more "wild" areas to hikers. The majority of tourists, Marshall asserted, would receive "far more pleasure and aesthetic stimulation" from a loop parkway outside the boundaries of the park. Trees and streams next to the road "are too close to be enjoyed at 40 miles an hour" and would lose their value because of the road itself. Rather, motorists were better served by the long views, by "looking at more distant objects which go by more slowly and do not depend for their enjoyment on quiet detail." Arboreal conditions in the park allowed for rewilding in large parts, as some 85 percent of its area had been logged by the 1930s. As a preparation for the scenic drive, residents of Cades Cove and other communities were forced out of the new park.[152]

Landscapes on the Blue Ridge Parkway needed restoration and presentation in the eyes of the planners. Scenery was not simply found. It was the result of design and planning after removing residents and their farms. The planners' goal was to present "a museum of managed American countryside," as Abbott put it.[153] This museum was a 469-mile (755-kilometer) road corridor with overlooks, parking areas, and the goal of controlling the roadsides and vistas. Recent dwellings were demolished, but "pioneer" mills and cabins became part of the mutual exhibit of landscape and farming culture. Families that lived on the land were relocated, although forced resettlement on the Blue Ridge Parkway was not as "draconian" as in the Shenandoah and Great Smoky Mountains National Parks, according to one historian. For Shenandoah National Park, several thousand residents lost their homes. After de-

A stop sign adjacent to the Blue Ridge Parkway. Some of the tensions between local residents and the Blue Ridge Parkway arose from its classification as a scenic road, not as a regular highway. Common-carrier traffic was and is prohibited on the road, and abutting owners do not have the same rights of access as they do to regular roads. National Archives, College Park, Maryland, Special Media Archives, Still Picture Unit, 50-4916

cades of portraying the former mountain dwellers as backward, the Shenandoah Park administration since the 1990s has been responding to pressure from descendants of the dispossessed. The movie played in the main visitor center now tells a more inclusive story. A lobbying group has succeeded in building monuments to the evictions.[154]

The Blue Ridge Parkway embodies the tensions resulting from the juxtaposition of the attractive surface of a scenic road and the forcible removal of residents that was deemed necessary in order to fashion that road from their land. The Park Service proudly displayed eighteenth- and nineteenth-century mills to visitors, using the most recent terrestrial transportation technology of the time on smooth, pleasurably paved roads. One of the most

Blue Ridge Parkway, Fox Hunters Paradise parking overlook (1953). Images such as the one above and the one opposite proclaim placid harmony and portray an idealized union of humans, technology, and nature on the Blue Ridge Parkway. Beneath the glossy visual veneer, however, social and cultural conflicts abound. National Park Service, photograph by Abbie Rowe

visited sights on the parkway was a gristmill with an overshot water wheel, Mabry Mills in Virginia. Park Service planners were chagrined to realize that it dated from the early twentieth century but pointed out that it looked much older. They razed the owner's 1914 frame house and "replaced it with a log cabin trucked in from another county." While planners opted for a pre-industrial appearance of the landscape, they provided a modern road transporting travelers to these imagined times and places.[155]

Control over the past and its narratives went hand in hand with control over views. Abbott and the other planners were keen to expand the right-of-way. Buying up land on both sides of the roadway ensured that the Park Service could control fauna and vistas by blocking undesirable views and opening up scenic ones. Initial plans to make the area of the park 1,000 feet (305 meters) wide faltered because of cost. On average, about one hundred acres per mile of road ended up being purchased by the Park Service, with

Blue Ridge Parkway, grazing sheep, Rocky Knob area (1952). National Park Service, photograph by Abbie Rowe

enough flexibility to gain "maximum control of the scenic picture with reasonable taking."[156]

Purchasing the land from private owners proved to be one of the most intricate issues for parkway officials. Some landholders were glad to sell marginal land, others were reluctant, and a few refused. Since ultimate control of the land would rest with the federal government, many locals saw the planning and construction of the parkway as an intrusion and responded with hostility. It became utterly clear to them that the road was being built for metropolitan tourists, not for them. Its legal status, as a limited-access parkway rather than a regular highway, meant that owners of abutting property had no rights of access. Intersections with regular roads were few. The Park Service's strict policy of banning trucks and even ambulances on the parkway amounted to a regime of exclusion. Thus, the polished appearance of the road conceals decades of conflict. Such is the "duplicity of landscape," as the geographer Stephen Daniels asserts.[157]

Concealing and revealing belong to the repertoire of landscape architects. In the case of the Blue Ridge Parkway, they sculpted a visual narrative
of an isolated population of humans working within a rugged landscape.
The long, sweeping views from high above joined with the presentation of
rural simplicity to produce a marked contrast to mid-twentieth-century
cities and suburbs. The tourist road concealed its origins in an increasingly
strong central state and in mass-produced conveyances. Instead, visitors
encountered a trip back in time to a frozen past.

The road also belied its specificity as a product of political wrangling and
design compromises. The landscaping of the planned parkway in the Appalachians was politicized before the first spade touched the ground. How much
scenery the road should produce, how it should relate to the Appalachian
Trail, and who should reap its benefits in economic, social, and cultural terms
were all under debate. Activists, politicians, professionals, and boosters intervened. While planners and wilderness advocates denigrated a skyline road
as either harmful or boring, the sectionalist jockeying and political battles
resulted in general design parameters that preferred ridgetop locations over
flanklines or valleys. Given their professional leanings, landscape architects
worked to create variety. To mountain dwellers, such distinctions among
urban elites mattered less. For many of them, the federal road was a scenic
infrastructure without immediate local benefit. For those who lost their
homes, it became a landscape of loss, if not alienation.

Like some other New Deal programs, the Blue Ridge Parkway aimed to
change land use patterns. Through publications and the work of agrono-
mists, soil conservation was taught to local farmers. On its own land, the
Park Service restored worn-out lands by having Civilian Conservation Corps
men rework the soil and plant seedlings. One example is Cumberland Knob,
close to the border between Virginia and North Carolina, where work on the
parkway began in 1935. It was remade as a landscape fit for hiking. Yet soil
conservation mattered less to the National Park Service in the ensuing years,
as completion of the road and tourism development became more pressing
and financial restraints were growing, especially in the postwar period.[158]

With increasing institutional momentum, the Park Service weathered
disputes with landowners, as well as the ebbs and flows of congressional
funding. It took until 1987 to finish the road. The Blue Ridge Parkway was an
unmitigated success in terms of visitation numbers. By the late 1950s, some
5.5 million annual travelers spent time on the road, and it was the most

heavily visited unit of the national parks. National newspapers reported on the opening of new segments and promoted the kind of auto tourism offered by this highway. Perhaps more than any other road, this parkway has helped to turn scenic driving into an experience that would last not just for a few hours, but several days.[159]

As the longest parkway ever built in the United States, the road stands as a monument to the gasoline-dependent exploration of landscapes. While tightly choreographed, the views from the road guide visitors to landscapes perceived as timeless and unchanging. The Blue Ridge Parkway shows the imprints of central planning during the New Deal, a disregard for local populations, and the elevation of regional scenery. Its intricate political history shows how driving and scenery were contested during the 1930s. Before he began to arbitrate these issues, Harold Ickes announced in the *New York Times*: "If I had my way about national parks, I would create one without a road in it. I would have it impenetrable forever to automobiles, a place where man would not try to improve upon God."[160] But Ickes did not have his druthers, at least not during the New Deal. In the messy, complicated world of a liberal democracy, with tensions between states, and between states and the federal government; with popular enthusiasm for automobiles; with professional opportunities for landscape architects; with dissenting voices arguing for minority rights in the wilderness movement; and with business and economic pressures from road and car lobbies, the planning and design of the Blue Ridge Parkway bore visible traces of these tensions. Professional and political elites at the state and federal levels wrestled to reach a compromise, while local residents often received short shrift. If anything, planners saw them as impediments to creating their vision of an Appalachian touring landscape. In the end, the parkway could not assuage these tensions and different claims.

The Longest Scenic Road

The Blue Ridge Parkway turned out to be the apex of the scenic road movement in the United States. Many drivers and lobbies appreciated, argued for, and experienced landscape-oriented rides in other countries. No other central government, however, has spent as many cultural, political, and financial resources on national parkways as America.

In contrast, the German Alpine Road was born of dictatorial simplicity. No due process, no public discussion, no fundamental questioning accompanied this offshoot of the overbuilt and ostentatious autobahn program. The Alpine Road was a response to other European efforts to coax and channel car-based tourism in the mountains. But it was also more than that. It was to be a European parkway based on American models, with the goal of re-making landscapes and remaking society along racially defined consumerist lines. Alpine scenery was spun into a web of triumphalist access to a formerly remote terrain. On the ground, however, the road was never finished. After early, intense activity around Hitler's mountain retreat on the Obersalzberg, other stretches received less attention. Ultimately, war and genocide were the regime's main aims, not appreciation and use of scenery.

The mid-1930s were parkway moments in both countries. Local and regional plans, projects, and actual stretches of road had given these scenic infrastructures a place in the car-road system. Their utility as beacons of work-relief programs provided a rapid infusion of political willpower and money. As nationally sponsored roads, both the Blue Ridge Parkway and the German Alpine Road sought to elevate the driving experience—literally and figuratively—by choosing higher altitudes.

Such higher-level roads and the ideas behind them incorporated decades of commercial and aesthetic practices, including mechanical panoramas, observation towers, cable cars, and gondolas. Such mechanically produced and accessed vantage points were now the province of the automobile—or should be, according to tourism promoters and roadminded builders such as Browning in Raleigh, North Carolina, and Todt in Berlin, despite their ideological differences. They preferred altitudes because they could produce more scenery; looking, for them, was overlooking aided by cars and roads. Roadmindedness attained new elevations. Such a view from above also took inspiration from the new observational techniques proffered by airplanes and cinematography.[161]

Parkway practitioners sought to align their ideas about beauty with the possibilities of experiencing landscape from a fast-moving car on curated roads. Alternating between peaks and valleys appeared to give the highest scenic return for landscape architects. Their professional rise was closely tied to parkways, especially in the United States, and they sought to plan them as carefully composed ensembles of road and surroundings. In the 1930s, they also sought to restore the roadsides to more pleasing levels. These roadminded designers incorporated highways into their expanding professional agenda.

Finally, the extent of road construction in the United States and its expected corollary in Germany provoked cries for abstinence. Hikers and, in the American case, wilderness advocates proclaimed less developed areas to be more valuable than obviously anthropogenic landscapes with new layers of roads. The point was not to set them aside for no use at all, but rather to preserve them for people with the physical fitness and time to traverse them on foot.[162] Activists such as Marshall preferred bodily vigor over both professional design and over providing automotive access to the summits, but he also recognized the role of scenic roads in less sensitive locations.

The bugbears of road critics were visitors to scenic places who only left their cars briefly. Visiting Logan Pass in Montana's Glacier National Park in 1935, the conservationist Rosalie Edge observed some motorists who stayed seated in their automobiles and several who only stepped out for a few minutes. They ignored the pleading of the park ranger to follow him on a hike to Hidden Lake, "rushing on in their enjoyment of perpetual motion." One year later, the German conservationist Hans Schwenkel encountered a group of six young women on the summit of the Hornisgrinde, a mountain in the Black Forest. They stopped their car at an observation tower, climbed it while chatting, and briefly looked around. After exclaiming "O yes! Very nice!" in unison, the motorists drove off. Schwenkel noted that the visitors were American. He warned his fellow Germans behind the wheel: if they thought they would immerse themselves in nature by speeding through it and stopping at beautiful spots, they were not only wrong, but had adopted alien habits of recreation and consumption. Such attitudes would make Germans "Americanized."[163]

While Edge warned against an incomplete and accelerated visit on comfortable roads, Schwenkel's diatribe branded such practices as foreign to Germany and synonymous with the most motorized and parkway-friendly country at the time.[164] Both shared a disdain for effortless scenic touring. The history of these landscape treatments was entangled. In the mid-1930s, scenic roads enjoyed the blessing and resources of central governments, but also provoked counterreactions. These discordant voices in the chorus of roadmindedness would soon become louder.

4

Roads out of Place

The traveler had a choice, and he opted against the scenic road. While touring the American Southeast, a writer for the *New York Times* decided not to use the Blue Ridge Parkway. Its ridgetop location made driving on an overcast day less pleasurable: "Rain, low clouds and zero visibility forced us off the Parkway but the Interstate also produces some sweeping vistas and runs up and down the slopes of many of those beautiful, sharp-edged Blue Ridges." When this account of an Appalachian journey was written in the early 1970s, the touristic desire for roadside panoramas had not abated. However, as a multilane interstate highway at a lower elevation and with a less poetic designation, Interstate 81 apparently sufficed for scenic intake.[1]

In the postwar years, parkways and the idea of automotive access to scenery became more and more controversial. The elite disagreements between design professionals and wilderness advocates were joined by more mass-based movements. In infrastructural terms, interstate highways, with their emphasis on constant speed and predictability, competed with the parkway planners' predilections for variety and visual enjoyment at lower speeds and sometimes higher elevations. A new kind of roadmindedness took hold with unadorned interstates, and new groups of actors contributed to redefining it.

As the following pages will show, the eventual dominance of utilitarian interstate highways was far from preordained. Pressure groups such as trucking lobbies preferred them over parkways, which they decried as wasteful. Driving on interstates proved to be different than touring on parkways. While American scenic roads continued to expand, such roads began to disappear from guidebooks both in the United States and in Germany. In cities, highways became controversial when urban activists pushed back against the destruction of their livelihood. For environmentalists from the 1960s onward, cars and roads stood for everything that was wrong with modern society.

Unbuilt Parkways

Given these contrasts and the rapid expansion of interstate highways, it is easy to overlook the plans for extensive parkway systems developed during and after the 1930s. Understanding unbuilt scenic infrastructures and the reasons why they remained on the drawing boards helps to illuminate the varied history of roadmindedness. Several parkway plans circulated in the press and among politicians. One of the longest of these highways would have been built in the country's most populated section. On the heels of the planning for the Skyline Drive and the Blue Ridge Parkway, Park Service Associate Director Arthur Demaray and Chief Thomas MacDonald of the Bureau of Public Roads introduced the idea of a "national parkway of the interurban type" from Washington, DC, to Boston in 1935. It would bypass the cities, thus lowering the cost of purchasing property. With at least 200 feet (61 meters), its right-of-way would allow for control of the roadside. Not only would this 450-mile (724-kilometer) parkway put thousands of people to work, but it would also serve the automotive needs of an area with some 25 million people, they wrote: "In no other section of the country is such a parkway so greatly needed." This proposal incorporated existing roadways around New York, in particular the Taconic Parkway being built by the State of New York, extending the Bronx River Parkway northward by 100 miles (161 kilometers) into the eastern Hudson Valley; Gilmore Clarke was the main landscape architect. From 1925 onward, Franklin D. Roosevelt had been a major driving force for the Taconic. His preference for an elevated route was incorporated into the road's southern portions. By the time the proposal for the national parkway was written, about sixty miles (ninety-seven kilometers) of the parkway from the George Washington Bridge to Poughkeepsie were complete. This interurban parkway's conception was close to the townless highway advocated by MacKaye and Mumford years earlier, but it was also grounded in the Bureau of Public Road's basic approach of responding to existing traffic needs rather than creating them. While the State of New York finished the Taconic by 1963, the larger federal plans for a road from the District of Columbia to Boston did not leave the basic planning stage. Neither did a scenic highway crossing the length of Massachusetts, despite the governor's offer to donate the right-of-way to the federal government.[2]

An even more ambitious (but also inchoate) parkway proposal was born in Congress. A 20,000-mile (32,000-kilometer) parkway loop was to span the entire continental United States and connect almost all national parks. On the East Coast, the parkway with a two-hundred-feet-wide right-of-way would have joined Acadia National Park in Bar Harbor, Maine, with Miami and the Everglades. Starting in Chattanooga, Tennessee, a southern route leading to the West would link several national parks en route to California. Sequoia and Yosemite National Parks were part of a northern section to Mount Rainier in Washington. From Seattle, the road would turn eastward and continue all the way to Albany. This monumental parkway proposal was introduced by the chairman of the House Roads Committee and received some press coverage, but little attention from Ickes or Roosevelt.[3]

Swept up in the parkway fervor, the American Society of Civil Engineers suggested a national system of car-only roads for tourists to separate them from trucks on federal-aid highways. The engineers justified their call with the "vastly predominating percentage of passenger cars now used for touring or pleasure purposes outside of municipal limits." Rather than calling them parkways, the society preferred "tourways" and presented a committee of professionals and road boosters; the chairman's role was reserved for Jay Downer, the prominent engineer with the Westchester County Park Commission. These tourways did not materialize, but they indicate just how high parkway fever was.[4] A 1941 "Park and Recreational Land Plan" published by the National Park Service described a national parkway system "as a move toward restoration, to the car owner, of those returns in pleasant driving, to which his payment of a variety of special and general taxes fully entitles him."[5] In other words, automotive scenery had become a federally funded entitlement.

Parkways vs. Highways

Given this broad level of support for not just local or regional but national parkways, the turn toward highways for both cars and trucks and with less concern for landscaping needs explaining. Roadmindedness was transformed. The push for parkways was strong enough to provoke a reaction. With the construction of several hundreds of parkway miles and more in the offing, a countermovement emerged. Trucking organizations provided some of the shrillest voices, since their vehicles were banned from scenic infrastructures. A spokesperson for the American Trucking Association reminded his readers that public roads were to be just that: open to anyone. Throughout history, he opined, roads had been used primarily for commercial purposes. He ridiculed the Skyline Drive as a "ribbon-way to fairyland" and thundered: "If the sightseer becomes so inanely selfish as to attempt or desire to legislate the other user off the highway, he admits ignorance of the purpose for which roads were built." A pamphlet by a trailer company derided scenic roads banning trucks as "horizontal monuments to poor planning." Truckers had to fight for a place on the open road. They loathed roads that were dedicated to passenger cars, both on the drawing board and on the scene in the Northeast and on Park Service parkways.[6]

Before and during the New Deal, highway lobbies began to push for the construction of a nationwide limited-access road system. Groups such as the National Highway Users Conference, which represented manufacturers of cars and trucks, as well as oil companies, argued that gasoline taxes should only be used for improving car and truck traffic. They also sought to uphold the professional primacy of civil engineers. With the increased level of funding for roads during the New Deal, federal plans for nationwide highways rested on the assumption that more trucks and cars would circulate on the roads. They should catch up with traffic. This approach had been the one favored by the Bureau of Public Roads early on, but it stood in obvious contrast to the traffic-inducing policy of the Park Service. The fact that the bureau was working with the latter did not resolve the underlying tension.[7]

In a foundational book-length report from 1939 entitled *Toll Roads and Free Roads*, the Bureau of Public Roads argued for need-based superhighways connecting major urban centers. The bureau suggested building new, limited-access roads where the need was already greatest and most likely to increase, based on traffic counts. The inclusion of trucks was so obvious as to not even be discussed. The statistics collated by the engineers made it clear that most trips amounted to five miles or less. Long-distance traffic was practically absent: less than 3 percent of all drivers wanted to go farther than 100 miles (160 kilometers). Traffic was mostly urban in origin and direction. Therefore, urban bypasses made little sense and traffic was to be directed into the heart of the cities, according to the report. Using Baltimore as a case study, the report called for combining slum clearance with highway building. World War II, of course, halted such efforts, but the fateful aim for the heart

An imagined intersection in the "World of Tomorrow," New York World's Fair (1939). At the height of federal parkway construction, visitors to the New York World's Fair in 1939 flocked to the "Futurama" exhibit, sponsored by General Motors and designed by Norman Bel Geddes. Neither a landscape architect nor a civil engineer, the industrial designer embraced cars and roads as modern and fast means of moving about. This multilane highway overpass embodies his modernist expectations. Norman Bel Geddes, *Magic Motorways* (New York: Random House, 1940), 98. Used by permission. Copyright: The Edith Lutyens and Norman Bel Geddes Foundation, Inc.

of cities came to full force afterward. Only a half-decade after MacDonald's and Demaray's proposal for a national parkway from Washington to Boston bypassing the urban centers, this more consequential document proposed commercial highways targeting downtown areas. Their best design was "level straight." Landscape development was now demoted to planting low-maintenance roadside vegetation rather than being part of the design process from the beginning. Civil engineers would be in the driver's seat.[8]

The detached professional language drawn from statistics in *Toll Roads and Free Roads* had its exuberant, utopian counterpart in the 1939 World's Fair exhibit called "Futurama." Sponsored by General Motors and conceived by the industrial designer Norman Bel Geddes, the three-dimensional model of the "City of Tomorrow" showed gleaming skyscrapers and highways with several lanes and on several levels. There was no congestion. In this wildly popular exhibit, public transit had all but disappeared in favor of a car-based urban modernity. Visitors observed the phantasm while sitting on a moving conveyor belt. Futurama equated urbanism with automobility and movement in city and country with wide, limited-access highways. The hinterlands in this exhibit of some 35,000 acres were marked by massive freeways with up to fourteen lanes. While civil engineers in the employ of the Bureau of Public Roads disliked the utopianism, they welcomed the public attention to highways and traffic problems.[9]

Expanding German Roads

Such vigorous public debates over parkways and highways were absent from Germany, where the professional primacy of civil engineers for roads was never fundamentally questioned during the Nazi years, the haphazard involvement of landscape architects notwithstanding. Fritz Todt's elevated role in Nazi governance, and his expansionist visions, provoked occasional grumbling by hikers and conservationists, but no serious dissent. Against this background, Nazi planners envisioned an Alpine network of scenic roads with the German Alpine Road as its core. Even before Austria's annexation, an engineer noted the rise of auto hiking, cited American parkways as an inspiration, and suggested combining existing and future Alpine roads into a transborder system of scenic driving. Motorway enthusiasts, in the meantime, sought to expand German and Italian interstate highways into a European network under German domination.[10]

While postwar Europe developed extensive highway networks, they were planned and financed by individual countries. After Germany's liberation, the establishment of the Federal Republic, and the beginnings of European economic integration, new highway connections across borders emerged. In West Germany and elsewhere, multilane, limited-access highways enabled motorists and truckers to drive at constantly high speeds. The Federal Republic inherited a highway system built far ahead of demand. Construction of new stretches, therefore, did not start until the 1960s, for the most part. Trucks had already been allowed to use Nazi autobahns. Unsurprisingly, trucking lobbies became some of the most vocal voices for rebuilding, maintaining, and extending the autobahn in the Federal Republic. As historians have shown, their pressure groups—and those of car manufacturers, tire

The West German autobahn (1981). As a busy transportation corridor, the post-war autobahn received public support and financial resources. Politicians and lobbyists thought of the network as the individualistic counterpart to suppos-edly collectivist transportation in East Germany and the Soviet bloc in general. Bundesarchiv Bildarchiv B 145 Bild-F060322-0004

producers, and other automotive interests—aligned the highways with a modern, Western way of moving about. Trying to strip the autobahn of its Nazi heritage, they stressed the capitalist values of moving freight quickly, and the personal freedom of the road in a liberal democracy. These lobbies were highly effective, not the least because they contrasted West Germany's embrace of individual transportation and multilane highways with the sup-posedly collectivist transportation system in East Germany.[11] A Cold War rationale prevailed, with individually owned automobiles and expanding high-ways as markers of Western consumerism.

When the United States began the construction of its interstate highway system in 1956, the conflation of free societies with individual, car-based mobility was at a Cold War high.[12] After decades of planning and political battles over financing, Congress assented to an interstate highway law that year. The federal government agreed to reimburse states for 90 percent of the construction cost for the new superhighways crisscrossing the country. Originally planned for 41,000 miles (66,000 kilometers), the interstates were the largest public works program ever undertaken by the federal government to that date. Its main motivation was not upgrading military logistics. Truckers and motorist organizations had been some of its chief proponents. This system cemented the utilitarian approach to freeways and the primacy of commerce and fast movement. Instead of using roadbuilding as a response to the economic calamities of the Great Depression, postwar federal highways were born out of affluence and a desire to provide automotive infrastructures for an increasingly car-oriented society. By providing the means to make long-distance road trips feasible and easy, they contributed to the decline of passenger railroads and the rise of the automotive sector as a pillar of the economy. As the interstate system was debated, one prominent conservationist, Bernard DeVoto, went so far as to support these highways wholeheartedly. They were needed for circulation and movement in a restless society, DeVoto wrote, and he implied that conservation and highways were not contradictory, as long as they were spatially separated.[13]

In West Germany and the United States, multipurpose highways for trucks and passenger cars became the norm for long-distance roads in the second half of the twentieth century, not landscaped parkways. Intricate and detailed guidelines issued by federal governments mandated design features such as the thickness of the road surface, the width of the road, and the size and design of signs in order to ensure uniformity in all processes related to roadbuilding. The decades from 1950 to 1980 saw the most rapid and extensive roadbuilding period in both countries. Given their new status as Cold War allies, knowledge exchange between West Germany and the United States accelerated.[14]

Road infrastructures grew by leaps and bounds, were highly popular, received copious funding, and, at least initially, engendered no widespread disagreement. Building predictable, uniform, high-performance roads suitable for all weather conditions and in various geographic zones was the goal of the civil engineers designing them. Drivers had to get used to operating their vehicles at constantly high speeds, weaving in and out of traffic, and to preparing to exit the highway. One important engineering worry was that of ensuring sight distance at high speeds—that is, to allow drivers to slow down and hit the brakes without collision, should they encounter a traffic jam or stalled car in the roadway. Drivers' visual perceptions mattered, but more as matters of safety, less so in the realm of aesthetic enjoyment.[15]

The experience of parkways, in contrast, had been about scenery, variety,

You Have to Slow Down to 90 for the Curves

A typical example of Goodrich improvement in tires

IMAGINE a flat, double ribbon of highway stretched taut over valleys, slicing between hills, running straight through huge tunnels bored in mountains. The grades so slight you hardly know you're climbing. Not a single stoplight or a crossroad in 160 miles. No speed limit. Few curves—and all of those can be taken at 90 miles per hour.

It's a dream highway—a dream that has come true. It's the new Pennsylvania Turnpike between Pittsburgh and Harrisburg. Building this super-express highway was a herculean project—a fantastic story of engineering triumph. Considered a three or four-year job, it was completed in approximately 20 months, employing as many as 25,000 men and thousands of

trucks at one time. While one crew cut down a mountain, others tore tunnels through solid rock, still others worked feverishly erecting the 285 necessary bridges.

Rubber made this highway possible! Special Goodrich Earth Mover and Rock Service Tires developed by engineers of The B. F. Goodrich Company were used on hundreds of trucks and earth-moving vehicles. Some of this equipment was so large that it picked up 25 to 30 tons of earth and rock, placed it in 8-inch layers, and

rolled it—all in as short a time as 12 minutes. Without these new man-high tires making possible gigantic loads, the Pennsylvania Turnpike probably would not have been built!

The development of these special tires is typical of the research work done by Goodrich to give you better truck and bus tires. For every transportation problem Goodrich provides special tires designed to give the longest possible wear for that type of service. If you want to cut your tire costs, call the Goodrich man. Remember which, the name's Goodrich. The B. F. Goodrich Co., Akron, O.; Los Angeles, Calif.; Kitchener, Ontario.

Goodrich Silvertowns
FOR TRUCKS AND BUSES

PHILADELPHIA	HARRISBURG	PITTSBURGH
1955 Hunting Park Ave.	2nd and North Sts.	5740 Baum Blvd.

Advertisement in *Highway Builder*, September 1940. Built on parts of a former railroad right-of-way, the Pennsylvania Turnpike allowed for unusually high travel speeds when its first stretch opened in 1940. This advertisement for a tire company extols the virtues of speed. Courtesy of *Highway Builder* magazine and relatively low speeds. Their sensory qualities did not place them outside the realm of capitalist circulation, but their low speed marked them as different. In the United States, they still found institutional sponsors. In the 1950s, New Jersey commissioned Gilmore Clarke to design the 170-mile-long (274-kilometer-long) Garden State Parkway, a toll road from suburban New York to New Jersey beaches. Trucks were banned on some of its portions and remain so today. This parkway offered a much faster ride than the suburban parkways of the interwar period, but the attention to landscaping was still paramount.[16] Almost simultaneously, workers in the Garden State paved the lanes for the New Jersey Turnpike, a commercial thruway for trucks and cars. In the state's flat terrain, the straight, unadorned, multilane road became synonymous with rapid movement, if not the state itself.[17] New York's state parkways were not finished until the 1960s. On the federal level, the most persistent patron of parkways remained the National Park Service.

Commemorating the historically important trade and travel route between Nashville, Tennessee, and Natchez, Mississippi, the Natchez Trace Parkway was built between 1937 and 2005 at a length of 444 scenic miles. Parkway historian Davis notes the "less dramatic nature of the terrain" and the drawn-out planning debates and funding battles regarding the road. In a more populous area of the United States, the Baltimore-Washington Parkway was opened in 1954. It combined parkway aesthetics with the characteristics of a high-speed commuter route.[18]

However, a popular vote had already brought down one prominent parkway proposal. The voters of Vermont defeated a planned Green Mountain Parkway in a referendum in 1936. It would have run the length of the state, from Massachusetts to the Canadian border, as one of the New Deal's relief projects. The landscape architect designing the route for the National Park Service worked on a version of MacKaye's flankline road; Robert Marshall objected not to the road as such, but to its proximity to the Long Trail, a hiking trail already completed by 1930, that runs along the north-south spine of the Green Mountains. Voters rejected the idea for various reasons, among them the yet-to-be-decided role of tourism in Vermont, the state's Republican political orientation at the time, and the realization that a Park Service parkway would not have allowed roadside development.[19]

One of the most ambitious and detailed American parkway proposals was for a Mississippi River Parkway, which would follow the banks of the river for 2,000 miles (3,200 kilometers), from Minnesota to the Gulf of Mexico.

Parkway for the Mississippi: Suggested location before construction. If built, the Mississippi River Parkway would have been the longest scenic road in the United States. While it did not leave the planning stage, the plan for the riverine road was only one of the imagined extensions of the 1930s parkway boom. These before-and-after images portray a shoreline road hugging the banks of the Mississippi. Bureau of Public Roads, Department of Commerce, and Department of the Interior, National Park Service, "Parkway for the Mississippi," Washington, DC: Government Printing Office, 1951, 27

Parkway for the Mississippi: Suggested location after construction. Bureau of Public Roads, Department of Commerce, and Department of the Interior, National Park Service, "Parkway for the Mississippi," Washington, DC: Government Printing Office, 1951, 28

An elaborate postwar report by the Bureau of Public Roads and the National Park Service concluded that new construction would be too expensive, and a toll road unnecessary and impracticable. Therefore, the agencies suggested combining existing roads and bridges, upgrading them to parkway standards, and adding some newly constructed interconnections. The road would re-

Protest hike, C&O Canal, 1954. An activist US Supreme Court judge, William Douglas (in the middle ground, raising his hat), leads the protest hike along the banks of the derelict Chesapeake and Ohio Canal. As a result of protests like this one, the National Park Service abandoned its efforts to turn the former canal into a parkway. Chesapeake & Ohio Canal National Historical Park

ceive federal aid and be "designed expressly for tourist travel and to conserve and develop the recreational resources of the region." A large survey, undertaken by a hundred planners appointed by the ten river states, examined the banks and existing roads. Congress, however, was unwilling to fund the road upgrades. Today, a "Great River Road" exists—if only on road signs—and is promoted by the tourism agencies of the states.[20]

A Consequential Defeat

Shorter in length, but probably more consequential in infrastructural terms was the defeated parkway on the remains of the Chesapeake and Ohio (C&O) Canal. This unbuilt road matters greatly in the history of road-mindedness. The C&O Canal paralleled the unruly Potomac, connecting the tidewater region of Washington, DC, with the headwaters of the Ohio River and Pittsburgh, and thus opening up the West and increasing the movement of freight. Or so it had been conceived by George Washington. When construction began on the canal in 1828, another infrastructure project aiming West began to make its way, less than fifty miles (eighty kilometers) to the north: the Baltimore and Ohio Railroad, the country's first. The latter project triumphed, and the former languished. The C&O Canal ended in Cumberland, Maryland, rather than beyond the Allegheny Mountains. By the 1930s, the federal government had taken over the decrepit canal ditch, which had flooded several times. During the New Deal, African American Civilian Conservation Corps men created a recreational landscape with hiking paths out of some of its remains close to the District of Columbia.[21]

After World War II, the Army Corps of Engineers came up with plans for a series of fourteen dams, given the frequent flooding of the canal. The storage lakes would have inundated the Potomac and the adjacent canal. Hydroelectricity and flood control were the goals. Another branch of the federal government disagreed. Seeking to preempt the Corps of Engineers, the National Park Service suggested filling the canal and turning it into a parkway from Washington, DC, to Cumberland. Factions of Washington's elites, including the Washington Post, were intrigued by the Park Service's plan, since it would "enable more people to enjoy beauties now seen by very few." But some hikers were prominent and in opposition. In a publicity stunt, Supreme Court justice William O. Douglas challenged the newspaper's editorial board to an eight-day hike along the 185 miles (298 kilometers) of the towpath in 1954. The ramble would demonstrate the recreational quality of this infrastructure for those propelled by their legs, not their automobiles. Tellingly, Douglas called the man-made landscape "a wilderness area where man can be alone with his thoughts, a sanctuary where he can commune with God and nature, a place not yet marred by the roar of wheels and sound of horns." Park Service Director Conrad Wirth objected to the first point. He emphasized that the canal area had already been a "commercial trafficway" for three-quarters of a century. After a series of disputes, the desire for non-motorized recreation won out in the end. The C&O Parkway remained unbuilt. The towpaths running on both sides of the canal, as well as some locks, were restored for visitors, hikers, and bicyclists. By 1971, the flood-prone structure was declared a National Historic Site; its maintenance fell to the Park Service.[22]

The defeat of the canal-turned-parkway was a major blow to the idea of scenic automotive drives in a prominent location. A generation earlier, the suburban parkways outside of New York had introduced metropolitan parkways to a larger and generally receptive, if not enthusiastic, audience. By the 1950s and 1960s, some of the most prominent local elites in the nation's capital rejected a suburban scenic drive planned by the entity that had been most active in promoting them at the national level.

More generally speaking, however, consideration of scenery in the context of driving was not completely abandoned. An ambitious 1966 national program instigated by President Lyndon Johnson proposed a $4 billion program to upgrade existing highways and build new scenic roads and parkways, for a total length of 54,000 miles (87,000 kilometers). Proponents justified the expenditure by pointing out that almost 13 percent of all driving, as measured in miles, was for pleasure. However, no new scenic roadway construction ensued. Instead, Lady Bird Johnson lent her support to a Highway Beautification Bill, which targeted billboards alongside federal roads. Historians agree that it did not achieve its goals; billboards were there to stay. Scenic roads, though, are still within the purview of the federal government. Since 1991, the Department of Transportation has been awarding the sobriquet of "scenic byway" to already extant routes that have undergone a vetting process.[23]

Within its own confines, the National Park Service continued to promote roads. Parks became ever more popular as the Park Service welcomed motorists, both in its narratives and on the ground. Beauty spots were easily accessible to drivers and passengers. Growing numbers of vacation days for workers and employees, the rise in discretionary incomes, and the pervasiveness of automobiles for suburban Americans contributed to a run on parks. In the 1960s, the Park Service managed to parlay this popularity into budget increases. The program was entitled "Mission 66." Some of this money went into building more and faster roads in the parks. Architecturally speaking, the Park Service introduced modernist styles into building and road design and did away with some of the efforts to incorporate vernacular forms. Roadmindedness took on new forms and undergirded a massive expansion during Mission 66. Whether modernist or rustic, road infrastructure in parks continued to grow, to the point of utter automotive convenience. According to one study, one in six visitors to Great Smoky Mountains National Park never turned off their car engines in the 1980s, thus partaking in a completely sedentary scenic experience.[24]

Fundamentalist Critiques

The infrastructural activism of Mission 66 and the rising numbers of visitors also provoked a fundamentalist voice from within the National Park Service, arguing *against* roads and cars. Edward Abbey, a former park ranger whose advocacy for wilderness issues went hand in hand with nativist calls to limit immigration, famously warned of "industrial tourism" and advocated a ban on roads. For him, highways eased the movement of humans into parks and thereby threatened their "wild" character. Wilderness depended on the absence of infrastructures in Abbey's eyes: "The auto-motive combine has almost succeeded in strangling our cities; we need not let it also destroy our national parks," he admonished.[25] While unsuccessful in his efforts to minimize roadbuilding in the parks, Abbey managed to popularize a counterview on roads in parks: rather than being artistically acclaimed structures that provided access, they were destructive impositions. Abbey's moralizing and vibrant writing drew attention to the signaling role of roads. For wilderness advocates such as Abbey, their absence was a crucial marker. In lieu of considering design options or choosing locations, they regarded roads in and of themselves to be one of the clearest expressions of mass tourism run amok. Besides, the disallowance of roads became part of the political and legal definitions for wilderness areas. Opposition to the Blue Ridge Parkway had pushed some activists to form the Wilderness Society in the mid-1930s. This small but influential group scored a major legislative victory with the passage of the Wilderness Act of 1964. In the areas protected by this act, resource use as well as any permanent structures, including roads, became banned.[26] This different stance is exemplified by one of the country's oldest environmental organizations, the Sierra Club. In the 1950s, it changed its attitude toward roads drastically. After supporting highways in the Sierra Nevada rhetorically and financially for the first half of the century, the club now opposed them loudly.[27]

The infrastructural boom years of Mission 66 provided updates for campgrounds, visitor centers, and utility services in the national parks. They also provided momentum for finishing the Blue Ridge Parkway. As Anne Whisnant writes, the Park Service was single-minded, if not always successful, in its pursuit of the road. It faced pressures from local developers and state politicians. The final holdup for completing the road was a long-standing conflict with a private company that owned a mountain peak called Grandfather Mountain, which was close to the intended routing. After adorning the mountain with a "mile-high" suspension bridge, the mountain's owners used the rhetoric of conservation to quibble with the design of the Blue Ridge Parkway. An expensive engineering compromise, in the form of a cantilevered bridge, called the Linn Cove viaduct, reduced the number of trees to be cut. By 1987, travelers could traverse the Blue Ridge Parkway on its entire length from Virginia into North Carolina.[28]

Continuing Regimes of Exclusion

While this road remained a sought-after destination for millions of visitors, its attraction was not equally distributed. By design, African Americans were planned to be segregated from white visitors at restrooms, campgrounds, and restaurants in the first few years of the parkway's history. In 1941, only about five thousand out of one million visitors to the Blue Ridge Parkway were Black.[29] After the war, segregated planning ended. By practice, however, the Blue Ridge Parkway was a public recreation area during the Jim Crow regime, which included spatial exclusion. Access to amenities of leisure, such as parks, was regimented and regulated in the American South and beyond. African American activists fought segregation in state parks through legal challenges. Even though formal segregation was declared illegal by the 1960s, participation in recreation continued to be decidedly unequal. On a deeper level, the idea of environmental leisure was itself racialized, as scholars have argued. Even in the absence of formal bans, practices and cultural norms of outdoor recreation continued to be associated with whiteness.[30]

In addition, steering an automobile to assert one's autonomy and automotive freedom were much easier to come by for white Americans than for African Americans. This was true not just in the South. Roadside businesses, such as gas stations or restaurants, might not cater to African Americans, necessitating careful planning of overland trips with guidebooks. The most famous one, the Green Book, listed establishments that offered services to them. Hostility, harassment by police, and (in the worst case) physical violence awaited unprepared Black drivers and passengers in some locales, as Gretchen Sorin documents.[31]

On scenic drives such as the Blue Ridge Parkway, these two regimes of uncertainty and inequality formed an uneasy admixture for African American motorists. Although it would be difficult to quantify how many of them used the recreational road in the postwar years, it stands to reason that they were few. The rising popularity of this road rested on white visitors, to a very large degree. In a rare piece of archival evidence from 1956, DeHaven Minkson, a prominent Black doctor from Philadelphia, expressed his reluctance to travel to the Blue Ridge Mountains, lest he would be denied dining and lodging. Rather than having to rely on the Green Book, Minkson sought to assert his rights as a citizen and as a motorist and wrote to the Department of the Interior. But such voices were rare in a geography of racialized exclusion.[32] Roadmindedness did not extend to everybody.

The Alpine Road after Hitler

Unlike the Blue Ridge Parkway, the German Alpine Road was never finished according to its original plans. Its incompleteness was the result and showcase of fractures. The political, economic, and moral ruin of Germany in 1945 provided the overarching rupture. In the postwar period, tensions between tourism and conservation, development and quietude, motorized access and sweat-driven solitude, as well as high- and low-altitude routes, came to the fore in much more pronounced and public ways than during the enforced acclaim of the Nazi years. Despite several lobbying efforts, the road links remained broken. Only after another, more recent, wave of lobbying did continuous signposts appear on the landscape in 2017 to elevate various stretches of road into a fully marketable tourist road.[33]

Immediately after the war, few things were further from the minds of Germans than tourism. Cities were in ruins, food was rationed, infrastructures still suffered from the war. Rather than thinking about getting away for leisure, most Germans were trying to scrape by. Millions of refugees and expellees needed shelter and nourishment. Yet, in Bavaria's Southeast, the Berchtesgaden tourist association was already preparing for a revival of tourism in early 1946. While it might seem paradoxical for them to have considered tourism issues at such a time, local leaders pointed out that recreation was the backbone of the economy in this part of the country. Once the general economy would rebound, Berchtesgaden would again become a favorite destination as the "Yellowstone Park of the Bavarian Alps."[34]

The reference to the United States, which had liberated this part of Germany less than a year earlier, and its national parks was not incidental—even if it was preposterous, given the sheer size of Yellowstone, which dwarfed the conservation areas in Germany. (Yellowstone would, in fact, cover one eighth of Bavaria.) Tourism managers realized that postwar tourism would be increasingly car-based and would require infrastructural expansion. The German Alpine Road, like many of the country's roads, lay in disrepair. Its most powerful sponsor, Hitler, had vanished along with his dictatorship. Instead of support from Berlin, the project of a tourist road in the German Alps now depended upon local and regional promoters. Instead of ready access to money, resources had to be negotiated through more democratic processes. Instead of carefully designed eye-catching flora and visual screens, spontaneous vegetation had crept up in some locations, and had overgrown isolated bridges and pieces of unfinished, crumbling roadbed.[35] To take just one example to illustrate the degree of contention: When a local member of the Bavarian state parliament suggested the construction of less than a mile of Alpine road (including an expensive tunnel) near the town of Schliersee in 1952, some members of a parliamentary committee growled that roads in Northern Bavaria were just as bad, and that their upkeep and renovation made more economic sense. While the state assembly ultimately approved Bavarian money for this stretch, the episode shows that intraregional tensions made it necessary to justify the road in the first place. The flow of money to the road was no longer a given.[36]

While roadmindedness thrived in general, the Alpine Road in and of itself, as well as its design, became controversial. Two years after the founding of the Federal Republic, a newspaper article ridiculed plans for expanding the road as premature and reminded its readers of the Third Reich support for the roads. The new federal government in Bonn was initially reluctant to fund the project. Only a few stretches could be justified to be built, and motorists could live with the gaps in coverage, since other roads were not far away, opined the writer. A suggested routing over the Kampenwand summit (5,476 feet, or 1,669 meters, above sea level) was now out of the question, and thoughts of the costs of doing so made state administrators "shudder," according to the paper.[37] The issue of elevation was directly related to cost: instead of connecting one mountain pass with another, as the Nazi plans had favored, the more economical and thus preferable postwar approach should be to circumvent peaks.[38]

Conflicts between the road administration and locals became more pronounced, too. A group of local farmers protested that the Alpine Road made it necessary to build fences to keep their grazing cows from crossing the

A teamster and a passenger in a horse-drawn cart travel on the German Alpine Road near Weiler (Allgäu) in 1952 when automobiles were rare. A tourist road on the scale that the Nazi dictatorship had envisioned became increasingly questionable during the postwar period. Critics of the project were now able to make themselves heard. Photograph by Willy Pragher (1952); copyright: Landesarchiv Baden-Württemberg, Staatsarchiv Freiburg, W 134 Nr. 021580b

road; they demanded that the state, and not they, should be responsible for building and maintaining the barriers. Some parliamentarians agreed.[39] Road planners in the Third Reich had prescribed wooden fences to control the view from the road in its totality. Local farmers, however, thought they were too "massive." Alwin Seifert had designed them, the Reich's administrators paid for them, but local farmers refused to accept the fences on their property, as they would then be responsible for their upkeep. Maintenance would have required an inordinate amount of lumber, in their opinion. The postwar federal road administration, indeed, preferred much less expensive (and less scenic) metal electric fencing to keep the cows at bay. While arguing whether Bonn or Munich would be responsible for financing the fences, all parties involved appeared to agree that simpler fences would, in the end, suffice.[40]

The Federal Alpine Road

Between 1952 and 1966, 90 miles (145 kilometers) of the German Alpine Road were rebuilt or built anew, which meant that 345 out of the originally planned 430 kilometers of the road were available. At the time, the Bavarian secretary of the interior indicated that completing the road would be difficult, given the high cost of construction at higher elevations and pressing needs elsewhere.[41] State planners responsible for transportation and infrastructures were keen to return to the interwar period's policies of counting traffic and prioritizing new construction projects according to needs, thus maintaining their department's expert status and its relative autonomy from politics.[42] Rather than bowing to political pressures, the department sought to rely on quantitative criteria. Measured by these yardsticks, the Alpine Road had not fared well before 1933, and administrators aimed to make it part of more comprehensive infrastructure planning after 1945.

The person most willing to lend his name to the Alpine Road after the war was Hans-Christoph Seebohm (1903–1967), the federal secretary of transportation. A mining engineer by training, he headed the Bonn department from 1949 to 1966 and made a name for himself through his shrill tirades on behalf of the German refugees from Central and Eastern Europe. He also loved to participate in opening ceremonies for the highways that were built from the late 1950s onward. His name is closely connected to the expansion of the autobahn network after the war. "No one thinks as highly of Hans-Christoph Seebohm as Hans-Christoph Seebohm," quipped the news magazine *Der Spiegel* on the occasion of the fifteenth anniversary of his position as secretary.[43] His management style was heavy-handed and his roadmindedness without question.

On one of his publicity tours, the secretary careened through Bavaria in the summer of 1959 to visit construction sites. East of Berchtesgaden, he ceremoniously opened a five-kilometer stretch of the Alpine Road to traffic and defended the overall road project, which was classified as a federal road. It was as economical as any other road, Seebohm declared, somewhat defensively. Tourists should be able to expect good roads when traveling. However, not every calm valley should be accessible through good roads. Seebohm enjoyed the public attention as a patron saint of the postwar Alpine Road, as one newspaper called him. However, he was less than sanguine in closed meetings. In the mid-1960s, local tourism boosters, including a monk from the well-visited Benedictine abbey of Ettal, formed a lobbying group to complete the Alpine Road. When they met with Seebohm, the secretary listed his prior support for the road, resulting in 83 miles (134 kilometers) of rebuilt or newly built road during his tenure. But he rankled the lobbyists by reminding them of the Nazi roots of the project and its avoidance of Austrian territory, which Seebohm deemed outdated, given the efforts at European integration. The Bavarians countered that the road still had its merits, even though the Nazis had sponsored it. "Apart from politics," opined the gingerly friar, who was also in charge of the lucrative Ettal brewery as cellarer, the "initiative back then" should be admired. Cutting off both religious and secular lobbyists, an agitated Seebohm insisted that the project was of regional, not national importance. A tourist road alone would not be eligible for federal financial support; such stretches might be built as toll roads. But if the road also served larger traffic functions, he would consider providing money. There was no dearth of resources for roads under Seebohm, but the Alpine Road did not make it to the top of the agenda, despite the Bavarian lobbying and opportune politics of memory.[44]

For one thing, the whiff of Hitler's personal love for the road continued to hang over the road like tailpipe emissions. One constituent derided the "phantasy project" of "Hitler's Alpine Road" in a letter to Seebohm's Federal Department of Transportation in Bonn. A centuries-old mountain pass which enthusiastic local Nazis had designated as the "Adolf-Hitler-Pass" was renamed to shed its odious associations. Not all infrastructure left over from the Nazi period was contested, but this scenic infrastructure emblematic of the charismatic dictator was. This does not mean that tourism in the Federal Republic was a radical departure from Nazi experiences, though. Even if they did not live up to their own promises, Nazi efforts to promote tourism in the *Kraft durch Freude* organization had popularized the notion of (racially grounded) extended vacations for wider swaths of society. Even more important, the memory of military deployments to Mediterranean or other European theaters of war was an often unspoken presence for postwar tourists. In the case of the town of Ruhpolding on the Alpine Road, a prominent package tour operator had begun, under Nazi rule, to offer vacations by chartered trains. During the Federal Republic, these greatly expanded trips brought tens of thousands of tourists by railroad to Bavaria. Gondolas and other Alpine infrastructures undergirded the rise of skiing, thus making the

Alps an all-season destination.[45] Bavaria attracted more tourists than any other German state, and tourism became an important pillar of a generally booming economy.

The Alpine Road contributed to and benefited from the spectacular growth of tourism in the mountains after the war. German and other tourists flocked to close-by destinations and increasingly used their own motorcycles and automobiles to do so. The Alpine Road certainly attracted more and more visitors to the Alps. But whether this was a development to welcome or to criticize was now an increasingly open question. The accessibility of the Alps, argued some, made them more vulnerable. Hikers could still find solace but not necessarily solitude in the mountains, given the number of tourists.

Elevation in the Crosshairs

Planners could no longer expect to be met with open arms when they discussed the Alpine Road. Apart from political battles over funding, organized opposition to building the highway on higher elevations arose. Conservationists were up in arms in 1965, when the newly formed lobbying group for the road presented a proposal for a mountaintop route; an extension leading from Linderhof Palace, near Ettal, to the town of Füssen was to be built over the Hochplatte summit (6,830 feet, or 2,082 meters, above sea level), in a nature reserve established just two years earlier. This was precisely the route that had served as the nucleus of the Alpine Road plans four decades earlier. Speaking for the proposal, the mayor of the tourist town of Oberammergau reckoned that bringing the road up to a mountain peak would create "one of the most beautiful and attractive" roads in the Alps. The mountains would "lend themselves" to such an elevated routing. A well-known hiking path would have been converted into an automotive route. According to the mayor, an Alpine Road worthy of its name must not be a "crawling valley" road; it needed to convey the beauty of high altitudes to compete with neighboring countries.[46]

The rhetoric of elevation was no longer paramount, however. The German Alpine Club, a group of hikers with tens of thousands of members, protested the project "not because we begrudge the car tourists an attractive road link, but because a nature reserve which is unique in Germany since it is untouched, extensive, and unique, can easily be endangered." Instead, the goal should be to preserve this landscape in its current state.[47] A more ecologically minded society noted that nature reserves in Germany were "ridiculously small" in comparison to the United States, the Soviet Union, Poland, or France. Therefore, one should not play fast and loose with the existing reserve.[48]

During the Nazi dictatorship, conservationists had not been decisively involved in Alpine Road planning. The growing postwar disenchantment with the production of windshield views rested, on one level, on the success of the interwar campaign for technified, consumable landscapes. Exactly because the Alpine Road had allowed busloads and carloads of tourists to experience the Alps while driving or being driven, the mountains had become less valuable in the eyes of middle-class urbanites who enjoyed summits for their remoteness. The memory of the Nazi past, so powerfully embodied in Hitler's personal mountain pass, meant that any lobbying for the road during the Federal Republic had to strip the project of its Nazi connotations and sponsorship.[49] As a result, the highly elevated road was not built; the only way to get to the Hochplatte today is by hiking. By the late 1980s, a local road builder summarized changing attitudes: "You can't impose on the landscape like that anymore."[50] The automotive avoidance of mountaintops had become commonplace.

German conservationists used the rhetoric and the legal weapons of conservation areas to fight ridgeline roads in the Federal Republic. Their reference to the supposedly "untouched" state of the Hochplatte summit evokes—but was not equivalent to—the American wilderness debate. Historically, German conservation had derived its strongest cultural powers from the idea of cultural landscapes. These were worthy of protection precisely because they were the result of both human and non-human forces, according to these voices. Criteria for selecting protected areas emphasized previous interactions between humans and environments. In contrast, the wilderness movement in the United States prized less developed landscapes, since they were closer to the perceived primeval state. One of the most important indicators for such pristine areas, their roadlessness, acquired legal force with the passage of the Wilderness Act.[51]

In the absence of such instruments, German conservationists pursued localized protests. As a result of postwar federal politics, shifting priorities, environmentalist critiques, and the all-too-prominent former sponsorship of the project by Hitler, the Alpine Road remained a patchwork of finished roads, detours, and privately owned toll roads. By 1978, a journalist judged it to be "likely one of the most beautiful, but definitely the most peculiar road in Europe." Almost randomly, the road appeared to veer in all directions; it had no signage on the ground and ranged from being a federally funded highway to a gravel path.[52] When the Alpine Road was half a century old in 1988, no entity stepped forward to celebrate the anniversary. Why not? A journalist for the liberal *Süddeutsche Zeitung* concluded that its incompleteness and the brief bout of attention lavished on the road by Hitler precluded a celebration. Still, he went on to describe how one would traverse Southern Bavaria from East to West using the existing stretches of the Alpine Road.[53] Rather than paying attention to the Nazi patronage, a newly founded umbrella organization for the Alpine Road paid homage to the 1927 proposals in 2002 and thus celebrated a seventy-fifth, apparently less innocuous, anniversary. However, this did not stop a newspaper from telling its readers that an eager local Nazi had renamed mountain features: in addition to the "Adolf-Hitler-Pass," two mountains near Bad Tölz were called "Hitler Mountain" and "Hindenburg Mount." The former had been adorned with a ten-meter high swastika lit by torches at night.[54]

The Disappearance of the German Alpine Road in Guidebooks

Needless to say, guidebooks for the Alpine Road avoided such images after 1945. Generally speaking, the degree to which landscapes were consumed and their manner of consumption are difficult to assess; the consumer's angle is notoriously absent from most sources. Still, printed tourist guides are valuable. In the case of the German Alpine Road, a survey of guidebooks shows the gradual but certain disappearance of the road itself as an object. The engineering work, the paved surface, receded into the background as the view from the road—not of the road—became more and more prominent.[55] In other words, visual expressions of roadmindedness were on the wane in guidebooks.

Without ever mentioning its former Nazi patronage, transportation secretary Seebohm gave the Alpine Road his blessing in a 1960 guidebook. In a preface, he stated (with some exaggeration) that the federal government had taken over the expansion of the road, had spent substantial amounts of money, and was on track to do so in the future. The guidebook, then, became a political document calling for the completion of the road and creating a continuity to its prewar sponsors without ever mentioning them. The guidebook's author, Hans Schmithals, was an artist who had already praised the first stretches of the road in a Nazi guidebook from 1936. Now, he celebrated the federal Alpine Road. Describing towns and sights along the route, the travelogue remarked on the inchoate character of the road.[56] A 1965 manual on driving in the Alps generously devoted twenty-three pages to the Alpine Road and its ancillary routes around Berchtesgaden, with several images of the road itself. The author noted, however, that the road was not completed, but drivers could patch together a ride through the Bavarian Alpine and sub-alpine regions by connecting the existing stretches of the Alpine Road with other roads.[57]

Realizing that the road would remain a torso, a 1970 guidebook cele-brated the sights that detours from the original plans would offer.[58] Some-what bashfully, the author noted that construction began "in late 1933"—every reader would get the reference. Disappointedly, the writer indicated that there was no prospect of its completion.[59] Pictures of the road are sparse in this book; cities, lakes, mountains, and other sights along its route dom-inate the visual impression.[60] Continuities were also personal. Wolf Strache, whose photographic career took off during the Nazi years, published propa-ganda books on the autobahn and worked as a military photographer for the Luftwaffe. By 1955, he celebrated the idea of "auto hiking" as if it was a nov-elty. Avoiding the term "German Alpine Road" and returning to the awk-ward interwar term "Queralpenstraße" (transversal Alpine road), he called it the "road of our dreams." Strache noted that the dream-like road remained thus, as it was never finished. Still, the existing stretches offered motorists "the enjoyment of complete floating, this unique combination of a deliber-ated technology and a landscape which still possesses the greatness and purity of the divine, this form of bliss that only modern man knows." For Strache, the road mattered because it overcame a supposedly apolitical ten-sion between technology and landscape in a harmonious way.[61]

By 1996, the Alpine Road had almost ceased to exist, pictorially speaking. An eponymous coffee-table book spent little time discussing the road itself, but celebrated the Southern Bavarian landscapes traversed by the (partially imaginary) road. Now, the focus was clearly not on the road itself, but on the way in which it connected travelers to sights and cities.[62] By 2003, another guidebook resigned itself to the fragmentary character of the road. Only then, apparently, was it possible to mention National Socialism by name. While noting the dictatorship's strong support for the Alpine Road, the author insisted that ideas for it predated the dictatorship, and that environmental concerns in the postwar period led to the road being unfinished. In the most Pollyannaish of interpretations, this was actually positive in the eyes of this writer: it would have been much worse if a comfortable, wide, multilane road had been built and attracted motorists from all over. The pictures made it clear that the attraction was not the road, but the Bavarian landscapes and folk festivals.[63]

The Disappearance of the Blue Ridge Parkway in Guidebooks

In the case of the Blue Ridge Parkway, guidebooks show a similar process of the pictorial disappearance of the road. The older and more extensive it became, the less there was of it to be seen in guidebooks. The road itself ceased to be a major attraction, but the view *from* the road never did.[64]

The oldest available guidebook, a National Park Service booklet from 1947, proudly showed an open, bending road on its cover. The road itself is only showcased in one more photograph out of eight; instead, wildflowers, picnic grounds, and the Mabry Mill site are depicted. Publications such as this booklet prepared tourists for their trips and planted a visual itinerary in their heads. They were also meant to be persuasive when it came to raising funds for the road's completion. From 1959 on, a quasi-official guide authored by an early park ranger named William Lord hit the bookstores along the parkway. It went through several editions. Without referring to the very public battles over routing during the 1930s, the guidebook presented the parkway as an apolitical achievement reconciling technology and landscape and making a romanticized version of Appalachia available to tourists. The 1982 edition of the guide, comprised of four volumes, featured the road itself on one of the four book covers and in two more images out of a total of fifty-eight. Not only had the road ceased to be a novelty; it was now a carrier to the attractions, not the attraction itself.[65]

A 1984 naturalist's guide celebrated the road as a respite from the usual entrapments of automotive landscapes. When driving its entire length, "you will travel a distance longer than that from Washington, DC, to Boston yet never pass a fast-food restaurant or drive-in movie theater. There is nowhere else in the eastern United States where a person can travel so far completely surrounded by trees." The guidebook, appropriately, devoted eleven of its twelve color plates to the fauna and flora to be encountered while driving, and one to an image of a car with dogwoods and redbuds in bloom. A late-twentieth-century publication with a short essay and seventy images included only three of the road itself. The proportion is similar for other coffee-table books.[66] In a 2002 pictorial guide, the Linn Cove viaduct receives a mention as an "engineering marvel," but the road at large has since clearly receded.[67] An international guidebook series summarized the parkway experience as "countless opportunities to be transported."[68]

When the first of these guidebooks was written, a continuous car-only highway without intersections over more than 400 miles (640 kilometers) was still novel. By the end of the century, both Germany and the United States had seen a rapid growth of their road network, and a road—no matter how well designed—became less of a destination. But both the Alps and the Appalachians retained their guidebook qualities as unchanging and majestic landscapes. Roads provided access to them and views of them. In other words, the notion of what a road is and what it does for sightseers had changed. One major reason for this refashioning can be found outside of touristic areas.

Freeway Revolts and Scenic Infrastructures

While the basic design features of the Alpine Road and the Blue Ridge Parkway changed little in the postwar period, roads in other settings, their shape, and routing became dramatically contested. This debate and its resulting cancellations or alterations of highway plans in turn greatly altered the understanding of roads and their relationship to urban and rural environments. A new generation of activists challenged roadmindedness, especially in its urban appearance.

The controversies over urban highways rested on a particular planning approach. American highway planners had been unwavering in their commitment to build roads from one city center to another, given their reliance on traffic counts. When state highway engineers began using federal dollars to plan interstate highways, they routed them through downtown areas and chose access routes that targeted corridors through the economic margins. Through a combination of looking for the least expensive land to purchase and their notion of "slum clearance," planners all too often targeted the homes of urban African Americans and recent immigrants with alacrity.

Such plans conveyed a technocratic confidence in planning for the common good and a professional myopia toward the social implications of a large-scale infrastructure. They engendered massive protests that historians have come to call "freeway revolts." In urban settings, roads became some of the most controversial large-scale technologies in the postwar era. As corollaries to the privately owned automobile, public roads featured in public debates. The American freeway revolt was more intense than the European one, the former being caught up in a Civil Rights struggle of asserting citizenship rights.

The planners' penchant for routing freeways through the heart of downtown areas and through established urban areas added to the depth of disagreements. To put it bluntly, urban freeways brought the infrastructural power and spatial reach of multilane roads into residents' backyards and changed their understanding of the urban environment. Robert Moses, who had used his political acumen to build parks and parkways in urban and suburban New York in the interwar years, became one of the most visible faces of postwar planning and construction of multilane, commercial highways in cities around the United States. For critics, his detached, utilitarian, and increasingly autocratic planning style personified an abrasive approach to public works. Throughput and flow were the key categories for Moses and for civil engineers aiming to move growing traffic expeditiously. Grassroots activists, on the other hand, saw an imposition on the urban fabric and fought against dispossession. Their neighborhoods were to be destroyed and their houses razed, with little benefit for themselves. In effect, their areas would become sacrifice zones. Given the intensity of the Civil Rights Movement and the fact that planners were almost always white male professionals, the freeway revolt was often seen through the prism of race. A slogan of the day from Washington, DC, castigated "white men's roads through black men's homes." Depending on organizational skills and political constellations, alliances cutting across class and racial lines were able to delay, change, or halt urban freeways in many American cities while other activists failed and roads were built. A generation of future politicians cut their teeth in freeway altercations.[69] These protests contributed to realigning relationships between experts and non-experts, between authority figures and the public, between politicians and citizens in general.

Among learned observers, the mood changed as well. Lewis Mumford, who had praised Moses's and Clarke's interurban parkways three decades earlier, railed against "the highway engineer's monstrous sacrifice of precious urban land to the accommodation of increasing traffic" and the "technocratic arrogance and ecological ignorance exhibited by current highway engineering" in the late 1960s. With the emotional depth of a jilted lover,

Mumford abandoned all hope when it came to cars and roads during this decade. His earlier optimism for creating restorative, well-planned carscapes had all but vanished.[70]

Given the tendency of civil engineers in Germany to have autobahn stretches bypass cities rather than traverse them, the European freeway revolt took on a more localized character. Arterial highways and urban ring roads, planned and built by municipalities, became the object of protest, lawsuits, and sometimes successful interventions. Criticism was more likely to be rooted in concerns for historic preservation than in the United States, and the altercations did not receive the same level of nationwide public attention.[71]

Among landscape architects in the United States, critiques of civil engineers and their work grew louder. Their former collaborators from the parkway era—landscape architects—were chagrined by having been excluded from the design process for the most part.[72] One particularly outspoken critic of the role of engineers was the landscape architect Lawrence Halprin. A Brooklynite, he trained at the then modernist Harvard Graduate School of Design and lived most of his life in San Francisco. Born in 1916, he no longer regarded parkways as a paragon of professional collaboration between civil engineers and landscape architects. His focus was on designing urban spaces and often on mitigating what he perceived to be the pernicious effects of urban freeways. An avowed modernist, he saw himself at the forefront of introducing new design elements into cities and landscapes.[73] Well known for his Franklin D. Roosevelt Memorial in Washington, DC, Halprin also designed Seattle's Freeway Park (opened in 1976), which showcased his artistic efforts to restore a ruptured cityscape after the construction of a wide traffic artery.[74]

In his 1966 book *Freeways*, Halprin clearly showed his admiration for well-designed rural highways and the revulsion he felt for urban freeways. From his point of view, the design of highways in the country was uncontroversial. He invoked the late-eighteenth-century landscape designer Humphry Repton's general plea for an "integration" of the road into the landscape.[75] To do so in the 1960s required "long, sinuous curvilinear patterns whose very calligraphy express the qualities of motion through space."[76] With the images presented, Halprin paid pictorial tribute to the Clarke generation of American parkways (the Mount Vernon parkway was shown) and the 1930s autobahn in Germany. For Halprin, there was little new to say on this.

The real issue, however, was the clash between urban centers and interstate highways. After professing his love for the "sensation of speed," the excitement of travel, and the beauty of freeways, Halprin criticized the "major disasters" that American cities had suffered. The problem was in designing freeways for the purpose of transportation alone, without regard for the urban fabric.[77] But he held out hope that highways and cities could be reconciled. Cities must be rebuilt, with freeways being integrated into them. Broad railroad corridors had disfigured urban centers and were moved underground later; learning from the railroad experience was key.[78] Halprin accorded blame to the roads themselves and their design.

Privately, he made it clear that single-minded civil engineers were the culprits. In a notebook entry, he castigated them harshly: "But goddammit the real trouble with highway design in this country is that it has been given over to a whole group of incompetent narrow gauge, limited, unknowing, inept people who are unable to deal or even understand the difficult sophisticated and complex problem. . . . Structurally they are babies, urban design-wise they don't have the foggiest notion of what we're talking about—on an aesthetic level they are boors on a planning level they don't even comprehend the problem."

While Halprin admitted to having written his remarks while riding on a plane and having consumed "several American Airlines martinis," one can assume that he spoke for some of his colleagues, if only more intemperately. His recommendation, as it was, consisted of "setting up educational procedures" for civil engineers so that they could fully appreciate the aesthetic and social dimensions of the problem.[79] Even though he shared few of the aesthetic tastes of the first generation of parkway designers, he joined them in claiming that civil engineers did not possess an adequate professional toolbox for tackling the problems of road design. Halprin reaffirmed the necessity of employing landscape architects and empowering them at the design level exactly when their influence on roads was waning.[80]

More generally speaking, these years saw the rise of a new generation of critics and planners who relegated parkways to an outdated effort to preserve unspoiled nature. Instead, design professionals such as the influential landscape critic John Brinckerhoff Jackson, and the postmodern architects Robert Venturi and Denise Scott Brown, embraced common and ordinary highways. Venturi and Brown famously embraced parking lots and symbolic architecture in their book *Learning from Las Vegas*. Jackson averred that a highway landscape "is beautiful when it offers freedom and community of experience." Gone were the efforts to immerse travelers in a sinuously progressing scenic corridor.[81] Jackson, especially, pursued a different version of roadmindedness and emphatically embraced highways rather than shunning them.

Regardless of such professional rivalries and changing views, popular sentiment, especially among urban protesters, began to turn against roads. When the Federal Highway Administration invited landscape architects, an architect, and a civil engineer to study the issue of urban freeways, it included Michael Rapuano, who had worked with Gilmore Clarke on parkway planning and had founded a consulting firm with him. A writer for the countercultural *Village Voice* in New York dismissed Rapuano as a "1930s highway philosopher," criticism that was sufficient to invalidate the entire project. For the freeway revolt generation, Rapuano and Clarke were simply discredited in the late 1960s, given their close association with Robert Moses.[82] Such anti-road sentiments culminated in the protests on the first Earth Day in 1970, an outpouring of environmental sentiment and crystallization point for the new mass-based environmental movement. According to historian Adam Rome, many Earth Day participants labeled the automobile as public enemy number one. "Cars were put on trial, buried, and hacked to pieces." In their eyes, cars and roads embodied what had gone wrong with society.[83] This was a fundamental critique. Rather than allowing nature lovers to access scenery or reconcile nature and technology, roads in and of themselves had become barriers to natural enjoyment and the functioning of ecosystems. The new environmentalism conjured catastrophes. In a list of "recent technological mistakes in the environment," the ecologist Barry Commoner cited the "maze of highways" and "hordes of automobiles" right after nuclear fallout.[84] Distinguishing between flankline and skyline roads appeared to be a quaint or dangerous exercise at a time when roads, as such, were the movement's enemy.

Parkways in a New Light

The view from the road changed dramatically during the post–World War II period. Newly built interstate highways expanded rapidly and dwarfed the interwar parkways in size and imaginative power. Plans for extended networks of parkways failed; focusing on scenery and excluding common-carrier traffic paled in comparison to moving cars and trucks quickly and in great numbers. Parkways such as the Blue Ridge and the German Alpine Road occupied niches, rather than serving as models for other infrastructures. During the interwar period, scenic infrastructures received general acclaim and were shaped by discussions among professional experts such as civil engineers and landscape architects. Their social and environmental costs received little attention in the general public. Utilitarian interstate highways from the 1950s onward enjoyed popular support and their networks grew rapidly. Increasingly, though, urban protesters and environmental activists began to question their location and design and, eventually, their fundamental purpose of enabling easier and faster movements for automobiles and trucks at the cost of destroying neighborhoods and of contributing to pollution.

In this process, the status and meaning of parkways and scenic infrastructures changed. As destinations in and of themselves, such roads faded. Their manicured and, in essence, didactic appearance could be understood as the "monotony of perfection."[85] Rather than being instructed to immerse themselves in the surrounding landscapes, drivers often preferred faster rides to partake of Cold War consumerism. In sum, the sheen of these duplicitous scenic roads was no longer as bright.

On regular highways and especially on high-speed interstates, however, road environments loomed larger than ever by the end of the twentieth century. A new roadscape emerged in this late-century version of roadmindedness. Drivers began to expect large signs legible from afar, rather than new landscape features evoking surprise at the end of a curve. From a psychological point of view, "it is a toddler's view of the world, a landscape of outsized, brightly colored objects and flashing lights, with harnesses and safety barriers that protect us as we exceed our own underdeveloped capabilities. What we see while driving is a visually impoverished view of the world." Sound barriers and jersey barriers provide visual effects of the more monotonous kind.[86] Instead of pursuing an educational version of roadmindedness, these roads simplified and standardized driving and vistas under the auspices of safety, not visual delight.[87]

In addition to the roads and their meanings, cars and their interiors changed as well. Braking and shifting gears demanded less effort or none at all; automatic transmissions became the norm in American automobiles. Automobile engines required less attention in general, allowing a process that scholars have described as cocooning. The interior of the car became a place of refuge and individual shelter. Car radios, once decried by MacKaye, became commonplace, as did climate control. High-altitude roads were no longer the only way to obtain cooler air during hot months. While parkways encouraged drivers and passengers to look outward and to see their ride as part of an environmentally grounded journey, the encapsulation of drivers in their automobiles allowed for a more inward-looking trip. The focus of postwar family trips became the collective experience of sights after arrival and less the trip itself.[88]

The kind of driver, then, who detoured from fast, predictable trips in favor of less utilized backroads was either searching for deeper diversion or trying to find distinction among homogeneity, and that driver had more time to do so. Books about these trips continue to find leisurely readers. Leisure, however, had become less and less of what roads were about.[89]

Epilogue

Landscape Taken for a Ride?

If latter-day versions of Ilya Ilf and Evgeny Petrov had visited the United States or Germany in the late twentieth century, they would have found much more extensive road systems than the ones they encountered in the 1930s. Rare are the places far from highways and interstates. They also would have found few observers who are enthusiastic about the environmentally redeeming qualities of roads. The optimism and high hopes that some groups and individuals placed in scenic drives and parkways had given way to a practice of mitigating environmental damage from emissions, noise, and habitat fragmentation.[1] The story of scenic roads, however, is not simply one of early embrace and late rejection.

Harmony and serene reconciliation between humans, technology, and the environment were the stated goals of these roads. Instead, planning, construction, and use revealed social and environmental cleavages. They exacerbated them in some cases. Paradoxically, the entangled histories of scenic roads, parkways, and commercial highways in the United States and Germany show how some actors highlighted the environmental dimensions of infrastructures and downplayed them at the same time.

It matters that one of the first significant planning and design features of the rise of cars and roads was a comprehensive (if exclusive) effort to cushion their spatial and environmental dimensions.[2] Aspirations were high and the claims of redemption were broad and deep. The scenic infrastructures discussed in this book were hardly a quaint interlude in the unfolding of the Motown cluster.

Roadmindedness had several dimensions. For the locals whose livelihoods they affected, very little about these roads was charming. Thousands of them were displaced in order to create vistas for visitors.³ These land-scapes built for mobile scenery furthered new social and cultural practices of automotive sightseeing. Some of their features became persistent. There can be little doubt that such roads contributed to making landscapes con-sumable for drivers and passengers en route and, in fact, helped to turn scenery into an automotive consumer item. Ilf and Petrov were onto some-thing: prospects became dispensed in almost the same manner as gaso-line in this emerging car-road complex. Associating the car ride with a view, whether bucolic or industrial, has been one of the legacies of the rise of self-propelled movement by automobile. This was hardly a necessary or straightforward development. The concern with automotive vistas has been an undercurrent of this book—whether they were sought after by scenically minded engineers or landscape architects, ignored by other planners, or actively countered by truckers and some planners. The issue was never re-solved. The strongest roadminded actors had access to and exercised state-sponsored power. In the process, scenic infrastructures demonstrated the might of central planning carried out by professional elites and backed up by governments. Road schemes received rhetorical and financial resources, while locals bore the brunt of disruption and displacement.

The embrace of scenic highways and the earnest belief in their curative powers was more than simply an overly optimistic embrace of a new tech-nological system. Historians have examined how some observers had high hopes for technologies such as telephones, aviation, or the internet when they first came into broader use. The connections they provided were to supersede barriers of creed or nationality, as proponents of technological optimism averred.⁴ In the case of scenic infrastructures, the expectations and plans were optimistic but more relational. In comparison to smoke-belching train engines and geometrically conceived train tracks, cars and landscaped roads were to remake transportation in an environmentally acceptable fash-ion. When pollution from heavy industry was omnipresent in industrial cit-ies and the railroads were the mobile version of heavy industry, automobiles and their infrastructures could be seen as nimble, smaller, less intrusive, and in fact restorative. The sins of heavy industry were to be expunged by scenic drives in landscapes of redress. In other words, scenic infrastructures were relational technologies with a keen, if socially exclusionary, eye toward their environmental dimensions.

Choosing cars and roads to cure the ills of transportation on rails is, of course, the exact opposite of today's environmentalist attitude toward transportation. Current proposals favor public transit and long-distance trains to help to overcome the dependency on cars and decrease pollution. Historically, however, these roles were reversed for some time. Historian Joel Tarr has observed how public health officials and sanitary engineers in large American cities welcomed automobiles in the early twentieth century. Introducing automobiles would reduce the number of urban horses used for deliveries and individual transportation. The millions of pounds of equine manure and the horse cadavers left in the streets were not just unsightly and smelly but provided breeding grounds for flies and other disease vectors. Cars would clean up cities and make them more healthful, or so the hope went.[5] Acting in parallel, landscaped roads were to have salutary effects as well by replacing coal-burning trains and regaining individualistic experiences of scenery.

At its peak, the parkway movement aimed for a perfectly choreographed experience of sightseeing, with drivers and passengers as participants in an individual sensory journey. At the same time, this landscape experience rested on the mass production of automobiles and their reliability. This was yet another tension inherent in roadmindedness. Fordism did not beget scenic roads, but it was one of the preconditions for their expansion. Even more so, the views were prescribed and framed by design professionals, leaving little to chance for the inhabitants of the car (unless they stopped and disembarked from their vehicles). This guidance and instruction, however, was made invisible. With some exaggeration (and unknowingly echoing Ilf and Petrov), wilderness advocate Bob Marshall likened the parkway experience to people moving on a conveyor belt through an art gallery.[6] He was disdainful but correct in pointing to the highly supervised nature of the experience.

In contrast, the rhetoric of the open road and of automotive freedom was pervasive on commercial highways, especially during the Cold War. The freedom to drive, to direct one's engine to any destination at any time, became ingrained as a measure of individual liberty, especially when compared to the collectivist means of transportation supposedly favored in the Communist world. As any motorist with a speeding ticket can attest, such parlance obscures the degree to which roads have become spaces of control, regulation, and surveillance. In popular culture, however, driving has retained its meaning as an escapist move.[7]

Also hidden to most drivers were the international dimensions of land-scaped roads. Civil engineers and landscape architects were engaged in ex-tensive transnational exchanges, but nationally planned and supported roads succeeded in the vernacular. In professional presentations and conversations, design features traveled easily across national borders. But the results on the ground were meant to reinforce allegedly national qualities and repre-sent the sponsorship of the nation-state.

The relational aspects of scenic infrastructures include professional, social, and cultural relationships across national borders. These roads (and protests against them) were the results of international conversations, not national cultures. Reports, scholarly papers, visitations, and the government-sponsored forums of the International Road Congress helped to create a constant flow of ideas, techniques, and relations. It would be both futile and pointless to look for one individual or country as the originator of scenic driv-ing. This book has presented an intertwined history. Countries and builders competed with and emulated each other. Centers of attention shifted. In the late nineteenth and very early twentieth century, Switzerland and other Alpine countries provided reference points and models of scenic infrastruc-tures for aspiring tourist regions elsewhere. For urban and suburban park-ways, the United States was the main display in the 1920s and early 1930s. Both Germany and the United States aimed to build extensive scenic highway systems from the mid-1930s onward with the Blue Ridge Parkway and the German Alpine Road as prime examples. At no time were roads or opposition to them merely isolated regional or national incidents. Roadmindedness and its contradictions traveled easily.

The earliest examples of scenic drives served as exhibits, as carefully arranged pictorial showcases of a new automotive access to nature in tune with middle-class, urban aesthetic approaches. Their rhetorical and archi-tectural distinction from railroads and the railroad journey was a social and cultural marker. Railroads with their prescribed views and schedules were for the masses, and newly found autonomy on roads and aesthetic gain were for the happy few who could afford automobiles.

The interwar embrace of consumerism resting on the mass production of cars in the United States appealed to Germans and raised the possibility of an automotive infrastructure for wider swaths of society, based on scenic enjoyment and environmental restoration in both countries. Roadmindedness took on new forms and sponsors. In the United States, parkways, a new class of roadways only for automobiles, emerged out of the urban parks movement. They represented an effort to restore property values and degraded landscapes, and out of a desire to clean up society and human dwellings in a form of eugenic environmentalism. The latter goal was particularly strong in New York. Parkways dotted the landscapes in other American cities and suburbs as well. In the eyes of their promoters, the chaotic, disjointed, and crassly commercial landscapes of public transportation and commercial thruways were to give way to orderly and environmentally stable roadscapes in line with middle-class ideas of respectability and social uplift. Cleansing landscapes went hand in hand with cleansing the populace in these cases. A concern for purity undergirded both efforts. While these suburban parkways received lots of attention, most of the roads that were built or upgraded in the 1920s, however, were federally sponsored commercial connectors for cars and trucks.

The New Deal in the United States and the Nazi regime's counterpart of public works fever created the political conditions for a wider embrace of the parkway idea. The Great Depression and political responses to it provided the juncture for not just expanding but reshaping roadmindedness. Central governments in Berlin and Washington, DC, employed metropolitan planners to bring forth national scenic roads in relatively remote settings with the Blue Ridge Parkway and the German Alpine Road. These roads were much longer and their implications were broader than previous examples. Fordist ideas of consumption and car ownership conjoined with extensive government intervention. Planners imposed their scenic visions of Appalachia and the Alps on local populations by using the powers of central planning. Locals were shunted aside. In the more open political system of the United States, the rise of these national scenic roads provoked a quest for roadless areas with wilderness as a shorthand for preserving areas from industrial civilization and its embodiment, the automobile and roads. Germany's dictatorship of roads allowed some grumbling on the sidelines, but no fundamental criticism. In the end, civil engineers and landscape architects with a view toward the moving image of mountain scenery dominated the design process in its fundamentals. Locals were excluded by design and practice.

Design features of scenic highways were remarkably similar, regardless of location. Curvilinear alignment, the cloaking and revealing of beauty spots by way of shrubs and trees, rest areas at prominent places, relatively low speeds, and sylvan or shoreline preferences had become part of an international design vocabulary by the 1930s. Both for the Blue Ridge Parkway and for the German Alpine Road, the rural and more extensive character of these highways resulted in roads being designed for the longer view from higher elevations. As national showpieces, they used altitude for visual and political effect.

The remarkable similarities in approach, vision, and execution of scenic infrastructures in Nazi Germany and the United States during the New Deal, however, do not mean that differences in political systems ceased to matter. Comparing, after all, is not equating.[8] Fordism and widely distributed cars were a reality in the United States by the late 1920s. For Nazi Germany, they remained an aspiration, as the regime attempted to create a racialized version of consumption. Planning processes differed; relative American transparency and debates contrasted with government by fiat in Germany. The simplicity of dictatorial planning, personified in Hitler and his road czar Todt, brooked no resistance for high-profile projects such as the Alpine Road. The roads were exclusionary as Germans defined as Jewish were legally banned from operating an automobile anywhere. While some hikers and preservationists were permitted to grumble in the remaining niches of publicity, there was no questioning the general undertaking. The status of state-backed experts, especially civil engineers, was not fundamentally challenged.

In contrast, the planning of the Blue Ridge Parkway rested on the political nature of expertise. The decision on whether to place the road in Tennessee or North Carolina in its southern part involved public hearings as well as public and backchannel lobbying by these two states. Scenic production rested on a political process involving state and federal actors. However, local residents were left with little access to these wrangles and often saw the road as an imposition on the land for the benefit of urban tourists rather than themselves, a "rich man's road."[9] For African Americans, the barriers for using national parks such as the Blue Ridge Parkway in the segregated South were formidable.[10] Spatial segregation by race continued after the end of legal, administrative exclusion; automotive recreation was not accessible to all. The extent of parkways and their wide footprint, both geographically and culturally, provoked a historically potent response in the form of the wilderness movement. For these critics, roadlessness rather than roadmindedness was the goal.

But the differences between the two countries go even deeper. The Nazi concern for the environment in general and scenic infrastructures in particular is best described as spasmodic. Given Hitler's and Todt's highly personal management style and, more importantly, the regime's ultimate priorities of war and genocide, it would be naive to look for a deliberate, sustained development of scenic policies or environmentalism during the Nazi period. By contrast, parkways played a specific role in the complex process of American environmental politics, as the rise of the organized wilderness movement shows.

In both countries, roadmindedness provoked professional rivalries. Civil engineers were the most obvious beneficiaries of the push to upgrade existing roads and build new ones during the automotive age. When it came to scenic roads, however, landscape architects aimed to assert their expertise as well. Their goal was to be involved in the initial design process rather than decorating a highway that others had planned. During the planning and construction of some American suburban parkways and of the National Park Service roads, they were able to achieve their goal. When interstate highways appeared on the drawing board, however, engineers tended to be in control. By contrast, Nazi Germany's chaotic system of governance and the power exercised by individuals such as Todt translated into uneven professional relationships. Civil engineers had the upper hand in the end. They managed to continue this status during the Federal Republic. While the Blue Ridge Parkway took several decades to build, the German Alpine Road itself was not completed as planned, not least because it bore the imprint of the dictatorship so visibly.

Institutions mattered, especially in the longer term. Both Nazi Germany and its partial successor, the democratic Federal Republic, did not have a counterpart to the National Park Service. More than any other historical actor, this agency has engendered scenic driving on roads designed for this purpose and translated it into governmental policy. Almost defensively, the Park Service's assistant director described it as "not being a construction agency, primarily" in 1944, when roadbuilding had been one of its most visible activities for some two decades.[11] The Park Service under Mather promoted car-based scenic tourism with bureaucratic fortitude and longevity. Roads such as the Blue Ridge Parkway and scenic automotive tourism as a common practice are the infrastructural and experiential practices of such institutional vigor. The cultural, political, and financial resources of this federal agency, even as it was one of many clamoring for support and money, promoted automotive scenic tourism for decades. When the bulk of traffic had migrated to infrastructures meant to facilitate volume and safety, the Park Service continued to hold on to its scenic roads out of institutional momentum and cultural preference. As a result, scenic roads and utilitarian roads sometimes run parallel in American spaces. No German equivalent of this spatial generosity exist.

The institutions and practices of the federal government also produced and contained within itself the wilderness movement. One of the most eloquent wilderness voices and parkway critics, Bob Marshall, was in the employ of the federal government while he provided an important intellectual and political impetus for the movement to set aside wilderness areas where roads were absent. Scenic roads and parkways did not beget the wilderness movement, but their proliferation provoked contestations over their design and location, which contributed to a legislative response in the form of the Wilderness Act.[12] Wilderness, of course, has deeper cultural and intellectual roots. During the historical moment of enthusiastic roadbuilding, however, a political movement emerged.

As a cultural concept, wilderness did not resonate in domestic German affairs, given its general embrace of man-made landscapes within its borders. Hikers and preservationists, however, were successful in stopping roads in sensitive spots such as Alpine summits after the end of the Nazi dictatorship. The reason for protecting these areas was not their supposedly untouched quality, but the fact that zones without infrastructural access had become rare. With growing postwar affluence, tourism expanded greatly, as did the desire to set aside areas that were only accessible on foot, if at all.

Both countries shared some similarities. After the war, during the greatest period of expansion for roads, parkways and the idea of the scenic drive did not wither, but they were overshadowed by more utilitarian highways. Driving habits fostered earlier, with their more immersive experience of landscapes, often made way for the faster reality of traversing mere territory and reaching one's destination quickly. Still, unlike the railroad journey with its unintended consequence of panoramic traveling, the twentieth-century driving experience continued to contain an element of slowing down for sights. The complex history of scenic infrastructures does not fall neatly into a narrative of rise and fall. Scenic highways were not displaced by utilitarian free-

"In Quiet Contemplation of the Scenery (In stiller Beschauung der Gegend)".
With a degree of irony, the caption to this snapshot from a private photo album
speaks of the "quiet contemplation of scenery." Automotive touring and sightseeing had become quotidian for many. Private photo album, not dated (1950s or
1960s), Stadtarchiv Munich

ways, just as cars and trucks have not replaced railroads. Rather, all of these
artifacts and systems were built on top of each other. They formed interconnected layers, not distinct areas.[13]

Automotive Scenery Today

The role performed by scenery has changed. Today, parents often
debate whether showing movies to children while driving is beneficial for
their offspring and for traffic safety. Citing the "mind-numbing boredom of
being strapped in a car seat for hours on end, with nothing to do but admire
'the scenery'" or "staring at mile after mile of Jersey walls," as one journalist
put it, Disney appears to win over the landscape, especially on long-distance
trips on busy interstates on the Eastern seaboard of the United States.[14]

What, then, is the legacy of these scenic roads and parkways? Did they
slow down traffic? Speed it up inadvertently? At first glance, such deliberately slower movements—less expeditious than fast trains and cars on

interstates—appear as precursors of the twenty-first-century slow move-ment in the realms of food and travel, with its middle-class emphasis on the local and the unaccelerated. As some observers have pointed out, this slow movement is born out of privilege. Pointedly, the geographer Tim Cresswell asks, "How bourgeois can you get? Who has the time and space to be slow by choice?" On their face, parkways would provide the underpinnings for such choices. Yet, they were also part of an accelerated world that brought met-ropolitan traffic to the farthest recesses of the nation-state with the help of new or extended infrastructures. Like the railroads that preceded them, they were part and parcel of a circulative system and of industrial moder-nity. They sought to provide a different version of modernity, not an escape from it.[15]

Telling the history of transportation and mobility in the twentieth cen-tury has often meant presenting a story about moving toward faster and faster means of transportation. Cultural historians have noticed and dis-cussed a contemporaneous sense of acceleration of life and the resulting responses, ranging from enthusiasm to fatigue and exhaustion. Yet, the his-torical moment of parkways raises questions about speed and its roles. With trains going ever faster and airplanes promising to break the boundaries be-tween heaven and earth, slower, scenic roads were much more than a retard-ing episode in the continuous march toward greater speeds. These highways show the interconnection of transport infrastructures: the purposefully slow movement on scenic roads was a counterbalance to the railroads, a re-sponse as much as an effort to find a more grounded movement. It was both deeply romantic and forward-looking as it sought to embrace a new technol-ogy and to mold it to reclaim what had been lost. Transport infrastructures are layered and connected, not distinct from each other, as this example shows.[16] Roadmindedness was an uneven and contested process related to other technologies, not a victory lap toward ever greater speed.

Slowing down was as modern as was acceleration. Attending to specific landscapes rather than an indistinct, blurred space to be traversed, showed environmental sensibility as much as it showed a selective embrace of scenic features derived from architectural and touristic conventions. The story of transportation and speed is incomplete without paying attention to de-celeration. The nineteenth-century trope of shrinking time and space as the result of new transportation technologies has had a long shelf life.[17] Park-ways, indeed, contributed to a shrinking of space, namely the perception that available space was shrinking and was getting more crowded with roads, rails, and other infrastructures. By the late twentieth century, the dominant perception was that available space had become less extensive, less malleable, and therefore more precious. The discussions over never-built parkways or the extensions of existing ones, the rejection of high-altitude roads in the 1960s and the 1970s, and the freeway revolt had been about the precarity and preciousness of spaces. In this new perception, space was now too valuable to use it for more roads. Space had shrunk, but not in the way that the designers of roads, in particular parkways, had imagined. Quite the opposite: instead of bringing drivers and passengers closer to nature, the extent of parkways and, especially, highways and interstates, had made more and more drivers, voters, and constituents feel that infrastructures estranged them from nature.

Finally, the interrelated, complex, and sometimes contradictory history of the automotive view from the road showcases how one generation's solutions to environmental questions can end up as the next generation's problem. The beautified, professionally supervised world of landscaped roads was intended to supplant the polluting railroad with its imposing infrastructures. By encouraging more driving and driving for the sake of driving, such ideas, plans, and their embodiments have helped to bring about a world of vastly increased gasoline-driven movement and one where non-motorized mobility has taken on new meanings. The view from the road has become a central aspect of our lives.

Notes

Introduction

1. Ilf and Petrov, *Ilf and Petrov's American Road Trip*, 5–7.

2. Rodgers, *Atlantic Crossings*; Nolan, *Transatlantic Century*; Dunlavy, *Politics and Industrialization*; Brown, *Plutopia*.

3. Lavenir, *La roue et le stylo*; Desportes, *Paysages en mouvement*; Hvattum, *Routes, Roads and Landscapes*; Mauch and Zeller, *World beyond the Windshield*; Merriman, *Driving Spaces*; Mukerji, *Impossible Engineering*; Stilgoe, "Roads, Highways, and Ecosystems."

4. Sims, "Transport"; United States Environmental Protection Agency, "Carbon Pollution from Transportation."

5. McNeill, *Something New*, 297. McNeill calls the automobile "a strong candidate for the title of most socially and environmentally consequential technology of the twentieth century" (310). For a somewhat different approach, see Koshar, "Organic Machines," especially 115.

6. As will become clear, I approach these topics by building upon the recent insights in the fields of mobility studies in the social sciences and the envirotech approach used by environmental historians and historians of technology. The literature on both is considerable and I cannot review it here at length. For examples and surveys, see Pritchard, "Toward an Environmental History of Technology"; Pritchard and Zimring, *Technology and the Environment*. My analysis is consonant with the notion of "envirotechnical ensembles" introduced in Pritchard, *Confluence*. For mobility studies, a few examples are Sheller and Urry, "The New Mobilities Paradigm"; Cresswell, "Towards a Politics of Mobility"; Divall, "Mobilities and Transport History"; Merriman/Pearce, *Mobility and the Humanities*.

7. Studies of infrastructures include van Laak, *Alles im Fluss*; Edwards, "Infrastructure and Modernity"; Larkin, "Politics and Poetics of Infrastructure"; Zeller, "Aiming for Control, Haunted by Its Failure."

8. For examples of the literature on aviation and airmindedness, see Fritzsche, *Nation of Fliers*; Syon, *Zeppelin!*; Palmer, *Dictatorship of the Air*; Pirie, "British Air Shows." For cultural histories of aviation, see Wohl, *Passion for Wings*; Wohl, *Spectacle of Flight*; Corn, *Winged Gospel*. For a discussion of "roadlessness" in the Soviet context, but with international references, see Siegelbaum, *Cars for Comrades*, ch. 4, "Roads," 125–72.

9. Solnit, *Wanderlust*, 90; Amato, *On Foot*; Zochert, *Walking in America*; Nicholson, *Lost Art of Walking*.

10. In these cases, most walkers and especially patrons would have been male.

11. Solnit, *Wanderlust*, 88, and ch. 6; Bermingham, *Landscape and Ideology*; Hunt, *Gardens and the Picturesque*; White, "Are You an Environmentalist?"

12. For urban walking and flaneurs, see Nye, *American Illuminations*, 41–42; Autry and Walkowitz, "Undoing the Flaneur," and the other papers in the fall 2012 issue of the *Radical History Review*.

13. Warner, *Streetcar Suburbs*. During the first decade of the twentieth century, about half of all urban commutes in Britain were on foot. By the end of the 1930s, the percentage had declined to 23 percent, but was still the most important mode, followed by bicycles (19 percent) and aboveground trains (18 percent). Pooley and Turnbull, "Modal Choice and Modal Change," 15; Amato, *On Foot*, 123–24, 150–51, 171.

14. Chamberlin, *On the Trail*, 26.

15. König, *Kulturgeschichte des Spaziergangs*; Burckhardt, *Why Is Landscape Beautiful?*

16. Worster, *Passion for Nature*, ch. 5, 118–48.

17. Solnit, *Wanderlust*, 178. Rosenzweig and Blackmar, *Park and the People*, 212–25. According to these authors, the "largest and most reliable group of visitors" during the first thirteen years of Central Park's existence arrived by horse or carriage.

18. Davis, *National Park Roads*, 20.

19. The literature on urban parks is rich, but I cannot do justice to it here. For a starting point, see Schuyler, *New Urban Landscape*.

20. Oettermann, *Panorama*; Crary, *Techniques of the Observer*. Circular panoramas were not as popular in the United States, where long canvases were more common. These canvases were unrolled on a stage while a narrator explained the scene. I am grateful to David Nye for bringing these distinctions to my attention.

21. Schmoll, "Aussichtsturm"; Speich, "Wissenschaftlicher und touristischer Blick"; Ekström, "Seeing from Above"; Bigg, "Panorama, or La nature a coup d'oeil."

22. Schivelbusch, *Railway Journey*; Revill, *Railway*, 36–56; Stilgoe, *Metropolitan Corridor*. For my analysis, I am excluding transportation on boats and ships on waterways. The perception of landscapes while traveling on these would merit further discussion. Nye, "Redefining the American Sublime."

23. As quoted in Schivelbusch, *Railway Journey*, 58.

24. Schivelbusch, *Railway Journey*, 64.

25. Freytag, "When the Railway Conquered the Garden." I am indebted to Anette Freytag for pointing me to this publication.

26. Andrews, "Made by Toile"; König, *Bahnen und Berge*; Vance, *North American Railroad*, 233.

27. John Muir, "The American Forests," *The Atlantic* 80 (August 1897), 154, as quoted in Nye, *America as Second Creation*, 194. Similarly, an early twentieth-century German traveler noted passing through dynamited forests in Canada: Holitscher, *Amerika heute und morgen*, 135–36. This travelogue appears to have been an inspiration for Franz Kafka's novel *Amerika*.

28. "To a striking extent, steelmaking in the United States was created for a single product: Bessemer steel rails." Misa, *Nation of Steel*, 42.

29. Andrews, "Made by Toile."

30. Nye, *America as Second Creation*, 193–94; Williams, *Americans and Their Forests*, 344–52; White, *Railroaded*.

31. The steeper grades resulting from early American building practices necessitated more powerful train engines. In Britain, tracks came closer to the ideal form of being straight, level, and smooth; engines were weaker. The term "minimalist infrastructure" is Robert C. Post's, in his review of Vance in *Railroad History*, no. 175 (Autumn 1996): 138.

32. Vance, *North American Railroad*, esp. 39.

33. For a few examples of the growing literature on tourism, see Walton, *Histories of Tourism*; Baranowski and Furlough, *Being Elsewhere*; Koshar, *German Travel Cultures*; Hachtmann, *Tourismus-Geschichte*.

34. Marx and Engels, *Manifesto of the Communist Party*, 477; Kocka, *Arbeiterleben und Arbeiterkultur*, 313–14; Hachtmann, *Tourismas-Geschichte*, 59–62; Roth, *Jahrhundert der Eisenbahn*, 185–95.

35. Orvar Löfgren, "Narrating the Tourist Experience," *Tourists and Tourism: A Reader*, ed. Sharon Bohn Gmelch (Long Grove, IL: Waveland Press, 2004), 106, as quoted in Sheller and Urry, "Places to Play," 4. Also, see Coleman and Crang, *Tourism*; Hachtmann, *Tourismus-Geschichte*; Spode, *Wie die Deutschen "Reiseweltmeister" wurden*. A discussion of bicycle-related travel and the experience of landscapes is outside of the scope of this book.

36. O'Connell, *Car and British Society*; Mom, *Atlantic Automobilism*; Houston, "A Proper Medium"; Wetterauer, *Lust an der Distanz*.

37. As quoted in Mom, "Civilized Adventure," 163.

38. Remarkably, internal combustion competed with steam-powered and electric propulsion well into the twentieth century. Kirsch, *Electric Vehicle*. On the environmental consequences of these technological choices, see McCarthy, *Auto Mania*. For repairs and maintenance, see Corn, *User Unfriendly*. Surveys on the history of automobility include Mom, *Atlantic Automobilism*; Ladd, *Autophobia*; Möser, *Geschichte des Autos*; Merki, *Der holprige Siegeszug*.

39. During the early phase of automobility up to World War I, motorists often encountered verbal and physical protests while traveling in the countryside. Ladd, *Autophobia*; Fraunholz, *Motorphobia*.

40. Barker, "German Centenary," 2. In 1960, 22 percent of American households did not have access to an automobile. This number decreased to 13 percent in 1980. Unsurprisingly, more urbanized areas showed higher percentages of households without cars. In New York State in 1990, 30 percent of all households did not own a car. Bureau of the Census, "Census Questionnaire Content."

41. In particular, the Ford Model T proved to be popular with farmers, who carried automobility in the United States to a much higher degree than in Europe. Kline, *Consumers in the Country*; Merki, *Der holprige Siegeszug*; Wells, *Car Country*. For a stimulating essay on the automobile as a multipurpose and multivalent vehicle, see Flonneau, *Cultures du Volant*, 8.

42. Lynd and Lynd, *Middletown*, 253, 255. The researchers also noted that walking for pleasure had become "practically extinct." For rural geographies, see Interrante, "You Can't Go to Town."

43. By the 1920s and 1930s, it was increasingly regulated and commercialized. Belasco, *Americans on the Road*.

44. The literature on consumption and consumerism has grown considerably; an important comparative and relational study is de Grazia, *Irresistible Empire*. For a global perspective, see Berghoff and Spiekermann, *Decoding Modern Consumer Societies*. On the United States, see Blaszczyk, *American Consumer Society* and the literature that she cites.

45. James, "Automobile and Recreation."

46. In addition to the literature cited in note 38, see Heitmann, *Automobile and American Life*, and the respective bibliographies of these books. For roads, see Dienel and Schiedt, *Die moderne Straße*.

47. Barker, "German Centenary."

48. Schmucki, *Traum vom Verkehrsfluss*, 60.

49. "Ausstattung privater Haushalte mit ausgewählten langlebigen Gebrauchsgütern am 1.1. des jeweiligen Jahres," spreadsheet provided by email, Statistisches Bundesamt, Wiesbaden, Germany, May 2017. The 1973 percentage of car-owning households was 55. Surveys regarding market saturation were not conducted prior to 1962. For the period from 1945 onward, I focus on the Federal Republic of Germany. For the Soviet bloc, see Siegelbaum, *The Socialist Car*.

50. Rieger, "From People's Car to New Beetle."

51. Möser, "Dark Side of 'Automobilism'"; Ladd, *Autophobia*; Fraunholz, *Motorphobia*.

52. Flik, *Von Ford Lernen?*

53. Nye, *America's Assembly Line*; de Grazia, *Irresistible Empire*. For the environmental dimensions of car production, see McCarthy, *Auto Mania*; Wells, *Car Country*.

Chapter 1. Roads to Nature

1. White House appointment book for Wednesday, June 7, 1916; letter from Wilson to G. E. Chamberlain, May 22, 1916, Letterbooks, ser. 3, vol. 29, p. 336, Papers of Woodrow Wilson, Library of Congress. Apparently, Oregon's senator Chamberlain had invited Wilson to participate. I would like to thank Bruce Kirby, Reference Librarian, Manuscript Division, Library of Congress, for his help with the Wilson papers. Three years earlier, Wilson had pushed another button in Washington, DC, thus setting off a dynamite blast in Panama to complete the construction of the Panama Canal. Carse, *Beyond the Big Ditch*, 113.

2. Lancaster, *Columbia: America's Great Highway*, 120. Unlike any of the other road testimonials, Lancaster's account is suffused with references to a Christian God and his creation.

3. As quoted in Ladd, *Autophobia*, 28.

4. In a speech to a Good Roads convention, Wilson also noted that roads were not "first of all" for pleasure travelers. Jakle and Sculle, *Motoring*, 33. For his change in opinion, see Wells, *Car Country*, 80. Exactly four days after the Oregon ceremony, Wilson signed the Federal-Aid Highway Act into law. Seely, *Building the American Highway System*, 41.

5. The road was completed in its entire length of 72 miles (116 kilometers) in 1922. The National Park Service calls this road the "the oldest scenic highway in the United States." http://www.nps.gov/history/hps/hli/currents/columbia/historic.htm, accessed February 17, 2010. Davis, *National Park Roads*, 90–91. Davis's comprehensive book examines all important parkways in the United States. For an environmental history of the Columbia River, and an important reflection on technology and the environment in general, see White, *Organic Machine*.

6. The road was to be at least twenty-four feet (seven meters) wide, "with extra width on all curves." No curve radius of less than 1 percent was allowed, and the maximum grade was 5 percent. Lancaster, *Columbia: America's Great Highway*, 114. These features would ensure a smooth ride for motorists.

7. Cornelius Vanderbilt, Jr., "Oregon Metropolis Marvel of Growth," *New York Times*, September 5, 1920, sec. 1, p. 13.

8. Lancaster, *Columbia: America's Great Highway*, 118. Also see Ochi, "Columbia River Highway."

9. In addition, Lancaster professed religiously grounded awe at the Columbia River gorge and aimed "not to mar what God had put there." Fahl, "S. C. Lancaster," 114.

10. The literature on the imperial gaze is considerable and not always applicable in the current context. A useful entryway is Pratt, *Imperial Eyes*. For a classificatory effort, see Urry, *Tourist Gaze*.

11. Sam Hill was a Portland booster who supported the expensive river highway even when few Oregonians owned automobiles. Fahl, "S. C. Lancaster," 114.

12. Examples include Koshar, "Organic Machines"; Seiler, *Republic of Drivers*; Ladd, *Autophobia*. For comparisons, see Mom, *Atlantic Automobilism*. For national examples, see Mauch and Zeller, *World beyond the Windshield*.

13. Lyth, "Supersonic."

14. Kern, *Culture of Time and Space*; Borscheid, *Tempo-Virus*; Benesch, "The Dynamic Sublime."

15. Barker and Gerhold, *Rise and Rise of Road Transport*; Guldi, *Roads to Power*; Wells, "Changing Nature of Country Roads"; Wells, *Car Country*; McShane and Tarr, *Horse in the City*; Greene, *Horses at Work*; Müller, "Beitrag des Chauseebaus."

16. Wells, *Car Country*; Merki, *Der holprige Siegeszug*; McCullough, *Old Wheelways*.

17. Wells, *Car Country*; Seely, *Building the American Highway System*.

18. Wells, *Car Country*.

19. Mom, "Roads without Rails." While Mom's paper is mostly interested in organizational aspects, my focus is more environmental.

20. Müller, "Beitrag des Chauseebaus." In what is today Germany, the road network grew from 15,500 to 71,500 miles (25,000 to 115,000 kilometers) between the mid-1830s and the mid-1870s. Kleinschmidt, *Technik und Wirtschaft*, 28; Lay, *Ways of the World*, 111–14; Schlimm, *Ordnungen des Verkehrs*.

21. Thomas H. MacDonald, "A Nation's Highways. Talk at Cordoba, Argentina, 9 September 1929," Box 10, Folder 43, William L. Mertz Transportation Collection, Collection #C0050, Special Collections and Archives, George Mason University Libraries, Fairfax, VA.

22. Seely, *Building the American Highway System*, 67.

23. Seely, *Building the American Highway System*, ch. 4; Rose, Seely, and Barrett, *Best Transportation System*, 40–43.

24. Jakle and Sculle, *Motoring*, 83.

25. An important institution in this regard is the American Association of State Highway Officials (AASHO), which was founded in 1914 with the "encouragement" of the American Automobile Association (AAA), as Seely writes. It played an important role in formulating technical standards that became mandatory for states accepting federal funds for roadbuilding. Seely, *Building the American Highway System*, 41–42.

26. Hokanson and Pappas, *Lincoln Highway*; Swift, *Big Roads*, 32–43, 99–101.

27. Weingroff, "Lincoln Highway."

28. Jakle and Sculle, *Motoring*, 43–53.

29. Weingroff, "From Names to Numbers." For the most famous of the numbered highways, see Krim, *Route 66*.

30. John Muir, *Our National Parks* (Boston: Houghton Mifflin, 1901), cited in Nash, *American Environment*, 72.

31. The English writer John Ruskin had famously wondered whether no nook of England would be "secure from rash assault" when a railway was being built in the scenic Lake District. For historical discussions, see Winter, *Secure from Rash Assault*; Ritvo, *Dawn of Green*.

32. *Nineteenth Annual Report of the American Scenic and Historic Preservation Society*, 1914, as quoted in Nash, *American Environment*, 76. Nash assumes that the author was George F. Kunz, the society's president and a jeweler in New York.

33. James, "The Automobile and Recreation."

34. For two examples of an extensive historiography on exchanges, see Rodgers, *Atlantic Crossings*; Nolan, *Transatlantic Century*.

35. Merriman, "Road Works."

36. Moraglio, "Transferring Technology"; Mom, *Atlantic Automobilism*, 572–81.

37. Lewis, "Statement." Lewis was Chief Engineer for the City of New York from 1902 to 1920. The problem of automobiles destroying road surfaces could either be addressed by controlling their speed, confining cars to special roads, or by reconstructing roads. Only the third option appeared feasible to Lewis. On this issue, see Wells, *Car Country*.

38. The only car-only roads Lewis could imagine were privately built racetracks such as the Long Island Motor Parkway, a forty-eight-mile (seventy-seven-kilometer) route financed by the auto-crazy great-grandson of Cornelius Vanderbilt. It was opened in 1908 and operated as a toll road after its incubation as a racetrack. The designation as a parkway is misleading, since speed was not a design objective for parkways at the time (see chapter two). One motivation for building the highway was the accidental death of a spectator at a 1906 race. A 1910 competition on the track watched by more than a quarter-million spectators left three drivers dead and twenty injured. "Alco Again Wins Vanderbilt Cup but Race's Death Toll Is High," *New York Times*, October 2, 1910, 21. It has been called the world's first limited-access roadway: Phil Patton, "A 100-Year-Old Dream: A Road Just for Cars," *New York Times*, October 9, 2008, AU10.

39. International Road Congress, *Third International Road Congress*, 280.

40. Gijs Mom, "Building an Infrastructure for the Automobile System: PIARC and Road Safety (1908–1938)," *Proceedings of the 23rd World Road Congress 2007*, 8, as quoted in Moraglio, "Transferring Technology," 18.

41. For pan-European exchanges, see Schipper, *Driving Europe*. For European highway meetings, see Heckmann-Strohkark, "Traum von einer europäischen Gemeinschaft"; Badenoch, "Touring between War and Peace."

42. "Address by Maj. Frederick C. Cook (Great Britain)," Permanent International Association, *Sixth International Road Congress*, 39–40, 39.

43. "Address by Mr. Thomas H. MacDonald, Secretary General of the Sixth International Road Congress," Permanent International Association, *Sixth International Road Congress*, 48–50, 50.

44. Quoted in Moraglio, "Transferring Technology," 22.

45. Gabriel, *Dem Auto eine Bahn*, 103; Davis, "Rise and Decline," 42–43.

46. Thomas H. MacDonald, "A Nation's Highways. Talk at Cordoba, Argentina, 9 September 1929," Box 10, Folder 43, William L. Mertz Transportation Collection, Collection #C0050, Special Collections and Archives, George Mason University Libraries, Fairfax, VA.

47. *The Roads and Railroads*, iii.

48. Post, *By Motor to the Golden Gate*, 246. In an act of masculinizing technology, Post left writing on "the subject of car equipment, driving suggestions, garage, and road notes" (241) to her son Edwin, who entitled his section "To the Man Who Drives" (243–50). On gendering technology and driving, see Oldenziel, *Making Technology Masculine*; Scharff, *Taking the Wheel*; Clarke, *Driving Women*; Clarsen, *Eat My Dust*; Ramsey, "Driven from the Public Sphere."

49. For a summary of such early twentieth-century trips, see McConnell and Pappas, *Coast to Coast by Automobile*. As these authors note, a major factor for these publicity trips was the promotion of American-built cars.

50. Post, *By Motor to the Golden Gate*, 3.

51. Mathieu, "Alpenwahrnehmung"; Tissot, "How Did the British"; Tissot, "Tourism in Austria and Switzerland"; Speich, "Mountains Made in Switzerland." Elitist travelers during the heyday of the Grand Tour from the sixteenth to the eighteenth century sometimes included a sojourn in the Swiss lowlands and regarded crossing the Alps on the way to Italy as a hazard, not a delight.

52. The literature on the topic is huge. Recent examples include the works by Jon Mathieu, especially his *History of the Alps*; Hansen, *Summits of Modern Man*; Keller, *Apostles of the Alps*.

53. As quoted in Frank, "Air Cure Town," 198.

54. At its opening in 1882, the Gotthard Tunnel was the world's longest. For a cultural history, see Schueler, *Materialising Identity*.

55. König, *Bahnen und Berge*.

56. Keller, *Apostles of the Alps*; Günther, *Alpine Quergänge*; Amstädter, *Alpinismus*.

57. Gottfried Eigner, *Naturpflege in Bayern* (Munich, 1908), 15–16, cited after Hölzl, "Naturschutz," 12. All translations are by the author, unless otherwise noted.

58. Quotes from Rudorff's philippic and a defense of such projects in *Zeitung des Vereins Deutscher Eisenbahn-Verwaltungen* 38, no. 1 (January 5, 1898): 14–15, available at https://books.google.com/books?id=SHYfAQAAMAAJ&pg=PA15&lpg=PA15&dq=rudorff:+Reisep%C3%B6bel&source=bl&ots=DZL-u44FuY&sig=ysIEZnXfA4am8PteUb_doDf-iYI&hl=en&sa=X&ei=8zBWVaDWEIKqyAT1xoGIDw&ved=0CB8Q6AEwAA#v=onepage&q=rudorff%3A%20Reisep%C3%B6bel&f=false, accessed October 1, 2019. On German conservationists, see, among several others, Lekan, *Imagining the Nation in Nature*.

59. Tissot, "Tourism in Austria and Switzerland."

60. Hans-Ulrich Schiedt, "Alpenstrassenfrage," *Wege und Geschichte* 1 (2002): 34–39.

61. Stephen, *Playground of Europe*, 369–71.

62. Stephen, *Playground of Europe*, 371.

63. Stephen, *Playground of Europe*, 373.

64. Stephen, *Playground of Europe*, 371–72.

65. Lavenir, *La roue et le stylo*, 172–73.

66. Furter, "Hintergrund des Alpendiskurses," 84–86. In the first decade of the twentieth century, crossing borders with an automobile was "a complex undertaking only feasible for well-connected people," as one historian puts it. Schipper, *Driving Europe*, 61. Just before and after World War I, the question of permits and insurance became somewhat easier to handle, but international travel was still very far from being commonplace.

67. Schiedt, "Entwicklung der Strasseninfrastruktur"; Schiedt, "Ausbau der Hauptstrassen."

68. Bierbaum, *Eine empfindsame Reise*, 179. One historian correctly notes that Bierbaum's trip was highly unusual and part of an effort to promote car ownership. Merki, *Der holprige Siegeszug*, 25–26. Given Bierbaum's desire to speak to a large audience and his popularity, however, his notions regarding landscape, speed, and travel are still noteworthy. Zeller, "Staging the Driving Experience."

69. Merrick, *Great Motor Highways*, 185. I am indebted to Georg Rigele for his observations on this book.

70. Abraham, *Motor Ways*, 217.

71. Merrick, *Great Motor Highways*, 188–89.

72. Freeston, *High-Roads of the Alps*, lists and compares these roads after feeling compelled to compare it to hiking: "The mountaineer is welcome to his extra thousands of feet; but no one can truthfully aver that nature as viewed from the roads themselves is not wondrously impressive and magnificently beautiful," 15. In a 1927 edition, Freeston notes approvingly the greater extent of and access to mountain roads and describes "motoring over Alpine carriage-roads" as an "invaluable intermediary between ordinary railway travel and the scaling of lofty peaks." Freeston, *Alps for the Motorist*, 4. An earlier section on "The Neglect of Road Travel" no longer existed in the 1927 edition.

73. Merrick, *Great Motor Highways*, xiv.

74. Merrick, *Great Motor Highways*, xv.

75. Merrick, *Great Motor Highways*, 102.

76. See, for example, this 1929 report about a Stelvio trip with a motorcycle and sidecar: O. Werner, "Unter den Firnen des Ortlers," *Motor-Tourist*, no. 2 (1929): 6–7. The author had received directions from a "young, friendly, German-speaking Fascist," 6.

77. Freeston, *High-Roads of the Alps*, 219–20, 227, 233 (quote).

78. Eduard Engler, "Auto-Reisen—einst und jetzt," *Motor-Tourist* (January 1913): 32–39, 36.

79. These accounts were also part of a competition between owners of internal-combustion engines and of electric cars. Kirsch, *Electric Vehicle*; Mom and Kirsch, "Technologies in Tension."

80. "Touristisches Preisausschreiben für Kraftfahrer: Wie sehen wir die Landschaft?" *Motor-Tourist*, no. 7 (March 29, 1929): 1.

81. Carl Graf Scapinelli, "Wie sehen wir die Landschaft? Zu unserem Preis-Ausschreiben!" *Motor-Tourist*, no. 8 (April 12, 1929): 1. Even though a few women were among them, the drivers were addressed as and conceived of as male.

82. Scapinelli, "Wie sehen wir die Landschaft?," 1. For analyses of the bird's-eye view,

see Dümpelmann, *Flights of Imagination*; Asendorf, *Super Constellation*. The classic analysis of the "God trick" is Haraway, "Situated Knowledges."

83. Zschokke, *Strasse in der vergessenen Landschaft*, 40.

84. Osmar Werner, "Eine Fahrt über die nördlichste Paßstraße Tirols. Der Flexen," *Motor-Tourist*, no. 8 (1929): 6–7, 7.

85. "Der DTC fährt in die Sächsische Schweiz," *Motor-Tourist*, no. 11 (1929): 2–3.

86. On road races, see Haslauer, "Kesselbergstraße"; Kotter, "Von Rennfahrern."

87. Christomannos, *Die neue Dolomitenstrasse*, 8.

88. Rigele, "Sommeralpen-Winteralpen," 128.

89. Keller, "The Mountains Roar."

90. Écomusée, *La route des grandes alpes*, 77, 92; Lavenir, *La roue et le stylo*, 212; Theo Gubler, "Die französische Alpenstraße," *Die Straße* 14 (1935): 512–14. During the Nazi dictatorship, a German observer noted the "peculiarly French triad" of building roads for general transportation, military, and touristic purposes simultaneously. Karl C. von Loesch, "Die Durchdringung der Alpen," in Finger und Wucher, *Die Alpenstraßen*, 9–13, 13.

91. Rigele, *Großglockner-Hochalpenstraße*; Rigele, *Wiener Höhenstraße*; Kreuzer, "Dieser zahlungskräftige Ausländer."

92. Buses and motorcycles, the mobile means of less affluent tourists, populated these roads especially after 1950.

93. C. E. M. Joad, *A Charter for Ramblers, or the Future of the Countryside* (London: Hutchinson, 1934), 28, as quoted in Merriman, *Driving Spaces*, 47.

94. Kaschuba, *Überwindung der Distanz*, 53.

95. Gabriel, *Dem Auto eine Bahn*, 23.

96. Speck, *Via Vita*.

97. Nye, "Redefining the American Sublime"; Lekan and Zeller, "Region, Scenery, and Power"; Schuyler, "Sanctified Landscape."

98. Runte, *National Parks*, 20–22.

99. Runte, *Trains of Discovery*.

100. Runte, *National Parks*.

101. National Register of Historic Places Inventory—Nomination Form, July 1978, available at http://pdfhost.focus.nps.gov/docs/NRHP/Text/79000253.pdf, accessed October 1, 2019.

102. Wilson, *Culture of Nature*, 100.

103. Barnett, "Drive-by Viewing."

104. Barnett, "Drive-by Viewing," 37; Spiro, *Defending the Master Race*.

105. Shaffer, *See America First*.

106. Wells, *See Europe If You Will*, 4.

107. Rothman, *Devil's Bargains*.

108. Belasco, *Americans on the Road*.

109. As quoted in Carr, *Wilderness by Design*, 146. For the relationships between railroads and the creation of National Parks, see Runte, *Trains of Discovery*.

110. Shankland, *Steve Mather*.

111. Quoted in Sellars, *Preserving Nature*, 28.

112. Sellars, *Preserving Nature*, 89.

113. Louter, *Windshield Wilderness*.

114. Carr, *Wilderness by Design*, 147, claims that the route was "officially designated" in 1921. It appears that no new roads were built for the park-to-park highway.

115. Mather to Katherine Louise Smith, Minneapolis, April 8, 1921, National Archives, College Park, MD, RG 79, Central Classified Files, Box 369, Part 7.

116. Mather to Katherine Louise Smith, Minneapolis, April 8, 1921, National Archives, College Park, MD, RG 79, Central Classified Files, Box 369, Part 7.

117. As quoted in Carr, *Wilderness by Design*, 147.

118. Seely, *Building the American Highway System*; Rose, Seely, and Barrett, *Best Transportation System*.

119. Seely, "Scientific Mystique."

120. Davis, "Rise and Decline."

121. Carr, *Wilderness by Design*, 174.

122. Schuyler, *New Urban Landscape*, 146. For an example of the extensive historiography on parks and urban planning, see Sies and Silver, *Planning the Twentieth-Century American City*.

123. McShane, *Down the Asphalt Path*.

124. Gudis, *Buyways*.

125. Robert Kingery, "Design and Layout of Roads to Serve the Needs of Metropolitan Areas," Permanent International Association of Road Congresses, Sixth Congress, Washington 1930, Second Section: Traffic and Administration, Sixth Question, Report 6-H, 6–13, 11. Kingery was general manager of the Chicago Regional Planning Association.

126. Belloc, *Road*, "author's introduction" (n.p.), 195.

127. For a discussion of the English example, see Merriman, *Driving Spaces*, 23–59.

Chapter 2. Roads to Power

1. Davis and Clarke, "Bridge-Building of the Sixth Engineers."

2. Seifert, *Leben für die Landschaft*, 27–28.

3. Zeller, "Molding the Landscape."

4. Reminiscences of Gilmore D. Clarke (1959), Oral History Archives at Columbia, Rare Book & Manuscript Library, Columbia University in the City of New York; *The Career of Gilmore D. Clarke, FASLA, FASCE, interviewed by Domenico Annese, FASLA* (Cambridge, MA: Video Services Center, Harvard University, 1977), Videotape at Loeb Design Library, Harvard Graduate School of Design, Cambridge, MA; Thomas W. Ennis, "Gilmore D. Clarke, 90, Is Dead: Designed Major Public Works," *New York Times*, August 10, 1982, B19; Annese, "Gilmore Clarke." At a 1962 conference in London, Clarke and Seifert presented their work as exemplars of highway landscaping. The two did not meet, however, as only Clarke attended in person. British Road Federation, *Landscaping of Motorways*.

5. According to one source, more landscape architects than members of other professions served in the NPS by the early 1950s and it was the largest employer of landscape architects in the United States, if not globally. Sellars, *Preserving Nature*, 184.

6. Historic American Engineering Record, Bronx River Parkway Reservation, HAER-NY 327, 2001, available at http://www.westchesterarchives.com/BRPR/Report_fr.html, accessed March 31, 2010; Weigold, *Pioneering in Parks and Parkways*; Davis, *National Park Roads*.

7. Davis, "Bronx River Parkway," 113.

8. On billboards, see Gudis, *Buyways*. In addition, hot dog stands and their decorations drew the ire of planners.

9. Jay Downer, "Grade Separation of Intersecting Highways," Permanent International Association of Road Congresses, Sixth Congress, Washington 1930, Second Section: Traffic and Administration, Sixth Question, Report 6-H, 17–20, quote 18. The quotation marks are Downer's. He also pointed out that the planners' goal was "reservational control over the right of way." Downer, 19. The total cost of the first fifteen miles (24 kilometers) of the Bronx River Parkway was "over $16.5 million." Mason, *Once and Future New York*, 182. The county expected to recoup the extravagant expenses by receiving higher taxes from real estate after the completion of the road. A contemporaneous study of parkway systems in Boston, Kansas City, and Westchester County could establish a "reasonably conclusive" causal relationship between parkways and the rise in real estate values only for Kansas City, mostly because of methodological issues. Nolen and Hubbard, *Parkways and Land Values*, 127. For the politics of roads in New York State, see Fein, *Paving the Way*.

10. Davis, "Rise and Decline"; Davis, "Mount Vernon Memorial Highway and the Evolution"; Davis, *National Park Roads*.

11. Mason, *Once and Future New York*, 190; Davis, "American Parkway as Colonial Revival Landscape," 149.

12. *The Career of Gilmore D. Clarke.*

13. Reminiscences of Gilmore D. Clarke, Columbia University.

14. Mason, *Once and Future New York*, 197. Davis also mentions Grant's involvement. Davis, "The American Parkway as Colonial Revival Landscape," 149–50. Grant also contributed to the 1922 "State Park Plan for New York," which proposed parks and parkways. La-Frank, "Real and Ideal Landscapes," 247. On Grant in general, see Spiro, *Defending the Master Race*.

15. For more on the complexities of ecological restoration, see Hall, *Earth Repair*; Hall, *Restoration and History*.

16. Stern, *Eugenic Nation*, 148.

17. Nolen and Hubbard, *Parkways and Land Values*. Of course, such an economic observation holds true for most urban parks as well.

18. A partial listing of these projects can be found in Annese, *Office of Clarke and Rapuano*. For an assessment of Clarke's design from an architectural perspective, see Campanella, "American Curves"; Zapatka, "American Parkways."

19. Caro, *Power Broker*.

20. Ballon and Jackson, *Robert Moses and the Modern City*. The quotation is from the introduction by the editors, p. 66.

21. Ballon and Jackson, *Robert Moses and the Modern City*.

22. Winner, "Do Artifacts Have Politics?" Winner relied on Caro in this paper. Woolgar and Cooper, "Do Artefacts Have Ambivalence?"; Joerges, "Do Politics Have Artefacts?"

23. Caro, *Power Broker*, 849.

24. Lewis Mumford, "The Sky Line: Bridges and Beaches," *The New Yorker*, July 17, 1937, reprinted in Wojtowicz, *Sidewalk Critic*, 185–89, quote p. 187.

25. Lewis, *Divided Highways*, 38.

26. Giedion, *Space, Time and Architecture*, 826, 831.

27. Reinberger, "Architecture in Motion."

28. Gandy, *Concrete and Clay*; LaFrank, "Real and Ideal Landscapes"; Radde, *Merritt Parkway*.

29. "States Are Planning Better Roadsides," *Better Roads*, February 1934, 22. The state highway commissioner also expected federal funds to become available for this parkway system.

30. Smith, *City of Parks*. I am grateful to Jennifer Alexander (University of Minnesota) for taking me on a driving tour of Minneapolis parkways.

31. Smith, *City of Parks*, 118. Wirth was the father of Conrad Wirth, director of the National Park Service from 1951 to 1974.

32. Cushman, "Environmental Therapy."

33. Michigan and Oregon were the first two. Cushman, "Environmental Therapy," 52–53.

34. Gubbels, *American Highways*, 2.

35. Gubbels, *American Highways*, 2. Gubbels also advocated abandoning roadside ditches in favor of wider and flatter roadsides and recommended hardy, native plants as the least expensive way of landscaping.

36. Cushman, "Environmental Therapy," 54–55; Gudis, *Buyways*.

37. Simonson, *Landscape Design*.

38. See the articles and discussions in *Landscape Architecture*.

39. Davis, "Mount Vernon Memorial Highway: Changing Conceptions."

40. Historical American Engineering Record, *George Washington Memorial Parkway*, 68.

41. Hughes, *American Prometheus*, 284.

42. Moraglio, *Driving Modernity*.

43. Zeller, *Driving Germany*, 50; Reitsam, *Reichsautobahn-Landschaften*.

44. König, *Kulturgeschichte des Spaziergangs*.

45. Ludwig, *Technik und Ingenieure*, 303.

46. Mierzejewski, *Most Valuable Asset*; Rose, Seely, and Barrett, *Best Transportation System*; Vance, *North American Railroad*; White, *Railroaded*.

47. "Von Wertach zum Adolf-Hitler-Paß. Eine der schönsten Strecken der Deutschen Alpenstraße im Bau," *Völkischer Beobachter* no. 170, June 19, 1935, clipping in Bayerisches Hauptstaatsarchiv MHIG 9216. The recent literature on the autobahn includes Schütz and Gruber, *Mythos Reichsautobahn*; Steininger, *Raum-Maschine Reichsautobahn*; Vahrenkamp, *German Autobahn*.

48. Schütz and Gruber, *Mythos Reichsautobahn*, 14.

49. Todt, "Der nordische Mensch," 395; Baranowski, *Strength through Joy*.

50. Hochstetter, *Motorisierung und "Volksgemeinschaft"*; Klösch, "The Great Auto Theft."

51. Klemperer, *Ich will Zeugnis ablegen*, 310–11, 329, 368, 370–71.

52. Rieger, *People's Car*; König, *Volkswagen, Volksempfänger*; König, "Adolf Hitler vs. Henry Ford."

53. *Dr. Todt: Berufung und Werk*.

54. Todt at the 1937 Reich Party Rally in Nuremberg, as cited in Schütz and Gruber, *Mythos Reichsautobahn*, 131.

55. Todt to Seifert, Deutsches Museum Archives, NL 133/56.

56. Zeller, *Driving Germany*, 53–55.

57. Zeller, *Driving Germany*; Gabriel, *Dem Auto eine Bahn*.

58. Cited after Zeller, *Straße, Bahn*, 83.

59. "Hitler Highway Best in Europe," *Pittsburgh Sun-Telegraph*, February 16, 1935, clipping in Bundesarchiv Berlin, R4601/1506.

60. Davis, "Rise and Decline," 43.

61. *The Career of Gilmore D. Clarke*.

Chapter 3. Roads in Place

1. Feuchtwanger, *Erfolg*, 88–93, vol. 1, ch. 11: "Der Justizminister fährt durch sein Land."

2. Stephan, "Feuchtwanger, Lion: Erfolg."

3. Daniels, "Marxism, Culture, and the Duplicity of Landscape."

4. Cutler, *Public Landscape*; Maher, *Nature's New Deal*; Woolner and Henderson, *FDR and the Environment*.

5. Maher, *Nature's New Deal*, 6.

6. Among historians, the role of the New Deal with regard to the power of the state is a vital topic of interest. For example, see the "AHR Exchange: On the 'Myth' of the 'Weak' American State," *American Historical Review* 115, no. 3 (2010): 766–800; Smith, *Building New Deal Liberalism*.

7. Patel, *Soldiers of Labor*; Patel, *The New Deal*; Patel, "Neuerfindung des Westens"; Blackbourn, *Conquest of Nature*; Garraty, "New Deal, National Socialism."

8. For recent treatments of Nazi economic history, see Tooze, *Wages of Destruction*; Schanetzky, *"Kanonen statt Butter."* On the environmentalist aspects of Nazism, see Blackbourn, *Conquest of Nature*; Brüggemeier, Cioc, and Zeller, eds., *How Green Were the Nazis?*; Uekötter, *Green and the Brown*. For a comparative history of some aspects of fascism, see Saraiva, *Fascist Pigs*.

9. Keller, *Apostles of the Alps*, 19. For inroads into a large literature, see Mathieu, *History of the Alps*; Bätzing, *Alpen*; Seidl, "Landwirtschaft."

10. Silver, *Mount Mitchell*; Davis, *Where There Are Mountains*.

11. For three examples of a vast historiography, see Pomeroy, *In Search of the Golden West*; Jakle, *Tourist*; Walton, *Histories of Tourism*.

12. Applegate, *Nation of Provincials*; Confino, *Nation as a Local Metaphor*; Lekan, *Imagining the Nation in Nature*; Blackbourn and Retallack, *Localism, Landscape, and the Ambiguities of Place*.

13. Seifert, "Trachtenbewegung."

14. Heißerer, *Ludwig II*, 89.

15. Schwarzenbach, "Eine ungewöhnliche Erbschaft."

16. "Schloss Neuschwanstein heute," https://www.neuschwanstein.de/deutsch/schloss /index.htm#:~:text=Neuschwanstein%20geh%C3%B6rt%20heute%20zu%20den,einen%20 einzigen%20Bewohner%20bestimmt%20waren.

17. Historians disagree about the design linkages: while Bright claims a direct relationship between Neuschwanstein and Sleeping Beauty Castle, Doss links the Disneyland designs to plans for Falkenstein castle, which was planned but never built during Ludwig's reign. Doss, "Making Imagination Safe," 183–4; Bright, *Disneyland*, 87.

18. Heißerer, *Ludwig II*, 130.

19. Bayerischer Rundfunk, *Vier Jahrhunderte Ringen.*

20. Shapiro, *Oberammergau*; Waddy, *Oberammergau in the Nazi Era.*

21. Lobenhofer-Hirschbold, "Fremdenverkehr"; Rosenbaum, *Bavarian Tourism*; Keller, *Apostles of the Alps*, 201–4; Kania-Schütz, *Deutsche Alpenstraße.* While tourism was still predominantly a summertime activity in the first half of the century, skiing grew in importance from the 1930s onward. Denning, *Skiing into Modernity.*

22. "Ein interessantes neues Projekt," *Füssener Blatt*, no. 74, March 30, 1929; clipping in Füssen City Archive, V 6 31–21; "Paßstraßenprojekt Füssen-Linderhof, Sonderabdruck aus dem Füssener Blatt vom 14. März 1930" and Wilhelm Hofer, "Bericht zur Begehung am 29. Mai [1930]", both in Füssen City Archive, V 6 31–20. Further correspondence is in V 6 31–21. The reference point for this idea was the Gaisberg road in Salzburg, which replaced a railroad to a vantage point overlooking the city by 1929. Rigele, *Großglockner-Hochalpenstraße*, 113–17. "Niederschrift der Sitzung des Arbeitsausschusses für die Paßstraße Füssen-Garmisch am 7. November 1930 im Verkehrsverbund München und Südbayern" and newspaper clipping "Das Projekt der Paßstraße Füssen-Garmisch," April 15, 1931 (newspaper unclear), both Staatsarchiv Munich, LRA 199267.

23. "Die Erschließung Südbayerns," *Münchner Neueste Nachrichten*, July 30, 1932, clipping in Füssen City Archive, V 6 31–20. Knorz was born in Northern Bavaria. On neo-natives, see Rothman, *Devil's Bargains.*

24. Biography August Knorz, 1999, provided to the author by the Town of Prien; Bundesarchiv Berlin, NSDAP-Gaukartei: Knorz, August; Bundesarchiv Berlin, VBS 1 (Parteikorrespondenz) VBS 1/1060015568; Sanitätsrat Dr. Knorz, "Grundlagen des Fremdenverkehrs in Südbayern," *Völkischer Beobachter. Bayernausgabe*, September 11/12, 1932, n.p.; Oberbürgermeister Dr. Samer, Füssen, "Warum die Alpenstraße notwendig ist," *Münchner Neueste Nachrichten*, no. 261, September 25, 1932, clipping in Füssen City Archive, V 6 31–20. While these tourism boosters focused on construction of a new infrastructure, the German postal service inaugurated a bus line all the way from Berchtesgaden to Lindau in 1931, using existing roadways. Comfortable buses with large windows transported tourists from Lake Constance to the Austrian border in two days. After the annexation of Austria in 1938, buses traveled all the way to Salzburg. World War II interrupted the scenic tours, but the tours recommenced in 1950 and did not cease until 1976. Baumann, "Auf den Spuren."

25. "Alpenstraße Lindau-Berchtesgaden. Das große Projekt macht Fortschritte," *Münchner Neueste Nachrichten*, September 22, 1932; "Eine bayerische Alpenstraße," *Münchner Zeitung* September 21, 1932; "Bau einer bayerischen Alpenstraße," *München-Augsburger Abendzeitung*, September 21, 1932; clippings in Füssen City Archive, V 6 31–20; "Betreff: Alpenstraße Berchtesgaden-Lindau," memo of meeting between Bezirksamt Traunstein and Chiemgauverkehrsverband, October 6, 1932, Staatsarchiv Munich, LRA 212383; "Die Queralpenstraße vom Bodensee zum Königssee. Vorschlag des Deutschen Touring-Clubs für die Streckenführung," *Der Motor-Tourist* 42, no. 11 (November 15, 1932): 2–3. In general, however, the Deutscher Touring Club, a middle-class motoring organization, remained unimpressed by the proposals for new roads and was among those voices calling for upgrading of roads rather than new construction. B. von Lengerke, "Erst Straßenausbau—dann Nur-Autostraßen," *Der Motor-Tourist* 43, no. 1 (January 15, 1933): 13.

26. "Für die Alpen-Querstraße," *Münchner Neueste Nachrichten*, January 13, 1933, clip-

ping in Füssen City Archive, V 6 31–20. Knorz, "Die bayerische Alpenstraße," n.d., Bayerisches Hauptstaatsarchiv StK 6950, claims that the motion carried on January 15 of 1933, but the full assembly did not meet on this day. "Bayerischer Landtag: Verhandlungen 1919–1933," http://geschichte.digitale-sammlungen.de/landtag1919/band/bsb00008701 accessed March 18, 2013.

27. "Der Landtag 1932–1933 (5. Wahlperiode)," https://www.bavariathek.bayern/medien -themen/portale/geschichte-des-bayerischen-parlaments/landtage-seit-1819.html, accessed January 5, 2022; Probst, *NSDAP im Bayerischen Landtag*, 108, 159.

28. Probst states that Nazi motions in late 1932 and early 1933 were increasingly geared toward consensus and drew support from the moderate parties. Probst, *NSDAP im Bayerischen Landtag*, 187.

29. "Das große Alpenstraßenprojekt," *Münchner Zeitung*, January 20, 1933; "Die bayerische Queralpenstraße," *Münchener Zeitung*, January 20, 1933, both clippings in Füssen City Archive, V 6 31–20. The second newspaper article mentions a meeting in the finance committee of the state assembly. The first clipping is most likely mislabeled and from the *Münchener Neueste Nachrichten* as the second includes the heading for the *Münchener Zeitung*. Vilbig (1874–1956) was a career administrator with an engineering background whose expertise on construction matters was welcome in the Nazi regime; he was promoted to "Ministerialdirektor" in 1937. Gelberg, "Oberste Baubehörde," 303–4; Lilla, "Vilbig, Josef." In his denazification proceedings, Vilbig presented himself as an expert who joined the party late and retired in 1939. Denazification file Josef Vilbig, Staatsarchiv Munich, Spruchkammern, Karton 1865.

30. "Auszug aus der Niederschrift der 19. Sitzung des Ausschusses für den Staatshaushalt vom 19. Januar 1933," Bayerisches Hauptstaatsarchiv, Landtag 14613.

31. Oberste Baubehörde, *Die Bayerischen Staatsstraßen*; Gall, "Straßen und Straßenverkehr."

32. Permanent International Association, *Sixth International Road Congress*; Oberste Baubehörde, *Die Bayerischen Staatsstraßen*, 16–17; Vilbig, "VI. Internationaler Straßenkongreß." In a statement for his denazification proceedings, Vilbig stated with some exaggeration that the 1925 report had been emulated in other German states and abroad. Denazification file Josef Vilbig, Staatsarchiv Munich, Spruchkammern, Karton 1865.

33. "Projekt der Queralpenstraße vom Bodensee zum Königssee," *Motor-Tourist* 43, no. 2 (February 11, 1933). The article contains some of Vilbig's comments from January 15 at the Bavarian state assembly.

34. "Auszug aus der Niederschrift," fol. 64.

35. "Auszug aus der Niederschrift," fol. 48.

36. "Auszug aus der Niederschrift," fol. 70. After all of this posturing, the parliamentary committee adopted both the Nazi and the Conservative motion after altering the former: instead of having the state pay for the road, special purpose associations should shoulder the financial burden.

37. Probst, *NSDAP im Bayerischen Landtag*, 199.

38. Verhandlungen des Bayerischen Landtags, 1. Sitzung vom 28. April 1933, S. 10, http://geschichte.digitale-sammlungen.de/landtag1919/seite/bsb00008706_00028 accessed March 18, 2013.

39. Sanitätsrat Dr. A. Knorz, "Die Bayerische Alpenstraße," n.d., Bayerisches Hauptstaats-archiv StK 6950, 5. Another copy of the memorandum was sent to Hermann Esser, Chief of the Office of the Prime Minister, in May. Knorz to Staatsminister Hermann Esser, May 13, 1933, Bayerisches Hauptstaatsarchiv, MHIG 9216. For a recent historical treatment of the planning history, see Kreuzer, "Die Deutsche Alpenstraße zwischen Natur."

40. "Die Queralpenstraße eine Reichsangelegenheit," *Berchtesgadener Anzeiger*, no. 75, March 30, 1933, clipping in Ramsau Town Archive; "Ein Paradies für Auto-Wanderer. Beim Vater des Alpenstraßen-Projekts," *Sonntag Morgen-Post. Nationalsozialistische Sonntagszei-tung*, no. 32, August 6, 1933, n.p.

41. Ruppmann, *Schrittmacher des Autobahnzeitalters*, 211. Frankfurt's mayor Ludwig Landmann, an ardent supporter of interwar autobahn plans, was expelled from his office in the spring of 1933 as the Nazis classified him as Jewish. Sternburg, *Ludwig Landmann*.

42. "Höhenstraße in den bayerischen Alpen," *Füssener Blatt*, May 6, 1933, clipping in Füssen City Archive, V 6 31–20.

43. "Zur Queralpenstraße eine Voralpenstraße" [no newspaper indicated, probably *All-gäuer Zeitung*], September 5, 1933, clipping in Füssen City Archive, V 6 31–20.

44. Memorandum, Ministerpräsident Sieber, September 13, 1933, Bayerisches Haupt-staatsarchiv, MHIG 9216.

45. "Alpenstraße Lindau-Berchtesgaden. Das gigantische Werk wird Wirklichkeit!," *Sonntag Morgen-Post. Nationalsozialistische Sonntagszeitung*, no. 44, October 29, 1933, 17.

46. Vilbig, "Die deutsche Alpenstraße von Lindau bis Berchtesgaden," July 1933, Bay-erisches Hauptstaatsarchiv StK 6950, 3.

47. Internal communications show that Vilbig scrambled to fund the design studies, but realized how urgent they were for the new rulers. Memorandum re: Alpenstraße, June 10, 1933, Bayerisches Hauptstaatsarchiv, MHIG 9216.

48. Ingenieur-Büro Widmann und Telorac (Kempten), "Queralpenstraße, Teilstrecke: Hohenschwangau-Schloss Linderhof, Variante zum Projekte vom Dezember 1932," Novem-ber 4, 1933, Füssen City Archive, V 6 31–20.

49. Heinz A. Küke, "Die Bedeutung der Alpenstraße für den deutschen Fremdenver-kehr," *Die Straße*, no. 7, 1935, 218–22.

50. "Alpenstraße Lindau-Berchtesgaden. Das gigantische Werk wird Wirklichkeit!," *Sonntag Morgen-Post. Nationalsozialistische Sonntagszeitung*, no. 44, October 29, 1933, 17–18.

51. The shorter one was 255 miles (410 kilometers) long and carried an estimated cost of 110 million Reichsmark; the longer one was 298 miles (480 kilometers) long and was es-timated to cost 134 million Reichsmark.

52. "Alpenstraße Lindau-Berchtesgaden. Das gigantische Werk wird Wirklichkeit!," *Sonntag Morgen-Post. Nationalsozialistische Sonntagszeitung*, no. 44, October 29, 1933, 1. For Todt's rhetorical combination of expertise and conviction, see Fritz Todt, Kampfbund der Deutschen Architekten und Ingenieure, Landesleitung Bayern, to Staatsminister Hermann Esser, June 9, 1933, Bayerisches Hauptstaatsarchiv, MHIG 9216; Seidler, *Fritz Todt*.

53. "Alpenstraße Lindau-Berchtesgaden. Das gigantische Werk wird Wirklichkeit!," *Sonntag Morgen-Post. Nationalsozialistische Sonntagszeitung*, no. 44, October 29, 1933, 17.

54. Behind the scenes, state administrators were scrambling to find money for the road in 1933, only to be relieved of their worries when the Reich government opened its coffers

for the former provincial project. The DDAC motoring club had jump-started the planning process with a subsidy of 3,000 Reichsmark after the Deutscher Touring-Club went bankrupt. Memorandum "Ministerratssitzung vom 29. November 1933," Bayerisches Hauptstaatsarchiv, MHIG 9216.

55. "Hauptlinie der Alpenstraße festgelegt," *Münchner Neueste Nachrichten*, no. 196, July 22, 1934, clipping in Bayerisches Hauptstaatsarchiv, MHIG 9216. According to the article, the goal was not to traverse ridges as highly as possible, but to provide the most attractive views.

56. Todt, "Anleitung für Entwurf und Bau der deutschen Alpenstraße," November 21, 1935, Bundesarchiv Berlin R4601/397.

57. Josef Fischer, "Entstehung, Linienführung und bauliche Ausgestaltung der Deutschen Alpenstraße," *Die Straße*, no. 7, 1935, 208–12; Todt to Siebert, April 30, 1934, Bayerisches Hauptstaatsarchiv StK 6947; "Höhenstraße quer über die bayerischen Alpen," December 31, 1932, newspaper clipping [newspaper unclear], Füssen City Archive, V 6 31–20; J. Telorac to Lord Mayor Samer, January 2, 1934; Widmann and Telorac to City Council, February 15, 1934, both Füssen City Archive, V 6 31–20; Fritz Todt to Lord Mayor Samer, Füssen, August 26, 1937, Füssen City Archive, V 6 31–20. For reservations on the Füssen-Linderhof road because of cost, see Staatsministerium des Innern, Weigmann, to Ingenieurbüro Widmann and Telorac, December 27, 1933, copy, Füssen City Archive, V 6 31–20 and Staatsministerium für Wirtschaft to Ministerpräsident, November 13, 1936, Bayerisches Hauptstaatsarchiv, StK 6999. However, an existing forest road to the east of Linderhof, which connected it with Oberammergau, was widened and upgraded in the summer of 1933 to allow motorists to travel to Ludwig's castle, to the chagrin of a local teamster who had moved tourists in his horse-drawn carriage on unimproved forest roads. Utschneider, *Oberammergau*, 135–38; "Gehorsamstes Gesuch des Lohnfuhrwerkbesitzers Karl Maderspacher" to Hitler, July 13, 1933; Bavarian Ministry for Economics to Maderspacher, August 21, 1933, both Bayerisches Hauptstaatsarchiv, MHIG 9216.

58. Chaussy, *Nachbar Hitler*; Mitchell, *Hitler's Mountain*. On Freud: Chaussy, *Nachbar Hitler*, 216–17.

59. Todt to Vilbig, July 12, 1935; Todt to Unterberger, April 17, 1936, both Bundesarchiv Berlin R4601/397. Todt to Unterberger, April 17, 1936, Bundesarchiv Berlin R4601/397.

60. Seifert to Staatliche Bauleitung für die deutsche Alpenstraße, March 6, 1942, Alwin Seifert papers 223, Technical University Munich-Weihenstephan, Chair for Landscape Architecture and Public Space; Seifert to Todt, May 6, 1935, Alwin Seifert papers, Deutsches Museum, archive, NL133/056. Seifert's care even included the design of pasture fences on the Obersalzberg. Seifert to Todt, September 1, 1936, Alwin Seifert papers, Deutsches Museum, archive, NL133/057.

61. Todt to Ministerialbauabteilung, Ministerium des Innern, Munich, December 1, 1934 (copy), Alwin Seifert papers, Deutsches Museum, archive, NL133/056.

62. Beierl, *Geschichte des Kehlsteins*, 24; Chaussy, *Nachbar Hitler*.

63. Beierl, *Geschichte des Kehlsteins*, 68.

64. For the Obersalzberg mystique sustained by the Nazi regime and the exploitation of locals on which it rested, see Chaussy, *Nachbar Hitler*; and Kershaw, *'Hitler Myth.'*

65. Michahelles, "Die Deutsche Alpenstraße," 1068.

66. König, "Roßfeld-Panoramastraße"; Schöner, *Rund um den Watzmann*, 72–75.

67. The Rossfeld as a film set, http://www.rossfeldpanoramastraße.de/informationen /das-rossfeld-als-filmkulisse/, accessed April 23, 2013.

68. Schöner, *Berchtesgadener Fremdenverkehrs-Chronik*, 103.

69. Biography August Knorz, provided by Town of Prien.

70. Fritz Todt, "Geleitwort," in Schmithals, *Deutsche Alpenstraße*, 5.

71. Otto Reismann, "Die Deutsche Alpenstraße," in Reismann, *Reichsautobahnen*, n.p.; Michahelles, "Die Deutsche Alpenstraße."

72. Koshar, "Organic Machines"; Koshar, "Germans at the Wheel."

73. Fischer, *Bayern links und rechts*, 14.

74. Todt, "Geleitwort"; J. Greiner, "Das Wunderwerk 'Deutsche Alpenstraße.' Die schönste und längste Alpenstraße ein Werk des Führers," *Völkischer Beobachter. Süddeutsche Ausgabe*, no. 243, August 31, 1935.

75. Todt, "Geleitwort," 5.

76. Hochstetter, *Motorisierung und "Volksgemeinschaft"*; Klösch, "The Great Auto Theft."

77. "Betreff: Straßenbau Weiler-Simmerberg-Oberstaufen," petition sent to Straßen- und Flussbauamt Kempten and to Dr. Siebert, Bavarian Governor, July 30, 1934, Bayerisches Hauptstaatsarchiv, StK 6947.

78. Staatsministerium des Innern to Deutscher Automobil-Club (DDAC), December 20, 1934, Bayerisches Hauptstaatsarchiv, StK 6947.

79. Josef Niggl, Gasthof "Zur Post" Irschenberg, to Esser, Bayerisches Hauptstaatsarchiv, MHIG 9216. For problems with winter maintenance, see Straßen- und Flußbauamt Traunstein to Bürgermeister, Inzell, January 23, 1938, Town of Inzell, Archive, File 631/10 and the correspondence in Staatsarchiv Augsburg, Straßen- und Flußbauamt Kempten 74. In the latter instance, Serbian prisoners of war were forced to shovel snow on the road.

80. Todt to Ministerpräsident Sieber, June 11, 1936, Bayerisches Hauptstaatsarchiv, StK 6947.

81. Straßenbauamt Kempten, *Die Jochstraße. Ein kleiner Ausflug in die Geschichte einer der schönsten Straßen Deutschlands* (Kempten: n.d., self-published), copy provided by Straßenbauamt Kempten; Erich Günther, "Befestigungen an Allgäuer Straßen," *Das schöne Allgäu*, no. 17, 1937, 273–4.

82. Landesausschuss für Naturpflege to Ministerium des Innern, August 8, 1933, Bayerisches Hauptstaatsarchiv, MHIG 9216.

83. Keller, *Apostles of the Alps*. Austrian Alpinists, however, did challenge the routing for the Großglockner road during its planning stage.

84. Todt to Oberregierungsrat Dr. Froschmaier, Berchtesgaden, Bezirksamt, November 26, 1935, Bundesarchiv Berlin R4601/397.

85. Todt to Bayerisches Staatsministerium des Innern, November 25, 1937, Bundesarchiv Berlin R4601/397.

86. For the Siemens plan, see Todt to Ministerialrat Unterberger, Bayerisches Staatsministerium des Innern (draft), May 11, 1938 Bundesarchiv Berlin R4601/402; on the Kraft durch Freude facility, see Straßen- und Flußbauamt Kempten to Generalinspektor für das deutsche Straßenwesen, June 13, 1939; Kraft durch Freude Zentralbüro, Steinwarz, to Generalinspektor für das deutsche Straßenwesen, June 23, 1939, both Bundesarchiv Berlin R4601/403.

87. Todt to Reichsführer SS, Verwaltungschef der SS, March 14, 1938, Bundesarchiv Berlin R4601/403.

88. Erhard Klein, *Der Schatten des KZ Dachau über dem Miesbacher Land* (Dachau: self-published, 2012), 22–44; Wrobel, "Sudelfeld."

89. Todt to Straßen- und Flußbauamt Traunstein and to Mr. Stöckel, Car Repair Shop, Ramsau, March 26, 1940, Bundesarchiv Berlin R4601/404. Todt's involvement likely stemmed from the proximity of the repair shop to his summer retreat in Hintersee, close to Ramsau.

90. Todt to Bayerisches Staatsministerium des Innern, November 25, 1937, Bundesarchiv Berlin R4601/401. For Todt's meddlesomeness, see Packheiser, "Die Deutsche Alpenstraße als NS-Prestigeobjekt."

91. Kraus, Naturschutzbeauftragter im Bezirk Füssen, to Bezirksamt Füssen, November 15, 1937, Staatsarchiv Augsburg, Bezirksamt Füssen 3591.

92. "Von Wertach zum Adolf-Hitler-Paß. Eine der schönsten Strecken der Deutschen Alpenstraße im Bau," *Völkischer Beobachter*, no. 170, June 19, 1935, clipping in Bayerisches Hauptstaatsarchiv, MHIG 9216.

93. Landesfremdenverkehrsverband München und Südbayern, "Errichtung eines Höhenwanderweges Lindau-Berchtesgaden," April 29, 1937, Staatsarchiv Augsburg, file Bezirksamt Kempten 106/7; "Ein Querwanderweg durch das Allgäu," *Völkischer Beobachter*, no. 208, July 26, 1936; "Von Ost nach West durch des Reiches Alpen. Laßt uns einen Höhenwanderweg von Berchtesgaden bis Lindau bauen!," *Münchner Neueste Nachrichten*, no. 130, May 14, 1937, clippings in Bayerisches Hauptstaatsarchiv MHIG 9216.

94. See the correspondence in Bayerisches Hauptstaatsarchiv MHIG 9219.

95. Kittel, *Maximiliansweg.*

96. Ludwig Siemer, "Die Alpenstraße als Mittel zur Hebung des Fremdenverkehrs im deutschen Alpengebiet," *Deutsche Alpenzeitung* 28 (1933): 235–40. An entire number of this journal was devoted to the Alpine Road, with the frontispice showing Hitler's Obersalzberg.

97. Otruba, Hitler's *"Tausend-Mark-Sperre."*

98. Merrick, *Great Motor Highways*, 103; Keller, *Apostles of the Alps*, 167–75.

99. Merrick, *Great Motor Highways*, 104. Merrick qualified the remark about window-dressing by claiming that the road was still "marvellous" and that he was "immensely grateful" for its existence. Merrick, *Great Motor Highways*, 104. Pagenstecher, "Automobilisierung."

100. Seifert to Todt, November 2, 1938, Alwin Seifert papers, Deutsches Museum, archive, NL133/058.

101. "Der Verkehr auf der Großglockner-Hochalpenstraße," *Die Straße*, no. 16, 1939, 540.

102. Lord Mayor of Lindau to Bayerischer Ministerpräsident, March 30, 1939; Fritz Todt to Bayerischer Ministerpräsident, April 21, 1939, Bayerisches Hauptstaatsarchiv StK 6950.

103. Blackbourn, *Conquest of Nature*, ch. 5, "Race and Reclamation," 251–310, quote 293.

104. Todt to Gubler, Basel, April 16, 1934, Bundesarchiv Berlin R4601/1107.

105. Casagrande, *Bodenmechanik.* According to his obituary, Leo Casagrande joined his brother and taught at Harvard University from 1953 to 1972. "Leo Casagrande, 87, Engineer and Educator," *New York Times*, October 27, 1990, 28.

106. For example, see the reports by Prof. Franz Kögler, "Straßen, Straßenbau und

Verkehr in den USA. Bericht über eine Studienreise im Juli 1936" and Emil Merkert's report on a meeting of the Society for Automotive Engineers in October 1937, both in Bundesarchiv Berlin R4601/1107. For examples of published reports, see Bruno Wehner, "Vergleichbares und Gegensätzliches bei deutschen und amerikanischen Straßen", *Die Straße*, no. 6 (1935): 180–81; Wolfgang Singer, "Parkstraßen in den Vereinigten Staaten von Nordamerika", *Die Straße*, no. 6 (1935): 175–78; Max-Erich Feuchtinger, "Kraftverkehr, Straße und Grünpolitik im Dienste der Erholung in Nordamerika," *Technisches Gemeindeblatt* 40 (1937): 51–54 and 78–80; Bruno Wehner, "Straße und Landschaft in den Vereinigten Staaten," *Zentralblatt der Bauverwaltung, vereinigt mit Zeitschrift für Bauwesen* 55, no. 13 (1935): 233–39; Erlenbach, "Straßeneindrücke von einer Reise in den USA," *Der Straßenbau* 28, no. 10 (May 15, 1937): 121–24; Bruno Wehner, "Parkstraßen in den Vereinigten Staaten von Nordamerika," *Der Straßenbau* 28, no. 11 (June 1, 1937): 133–38.

107. "Der Chef des Straßenwesens der USA bei Generalinspektor Dr. Todt," *Die Straße*, no. 14 (1938): 468. For MacDonald's assessment of German roadbuilding, see Seely, "Der amerikanische Blick."

108. Seely, "Der amerikanische Blick"; Seely, "Visions of American Highways," 269 (quote).

109. Clarke, *Architectural Forum*, May 1928, New York World's Fair Bulletin, June 1939, both in Bundesarchiv Berlin R4601/1441.

110. "Die Eröffnung des VII. Internationalen Straßenkongresses in München. Ansprache des Stellvertreters des Führers, Reichsministers Rudolf Hess," *Die Straße*, no. 1 (September 1934): 34–38; "Ansprache des Generalinspektors für das deutsche Straßenwesens Dr.-Ing. Fritz Todt," *Die Straße*, no. 1 (September 1934): 39–42; "Besichtigung der Reichsautobahn München-Landesgrenze und der deutschen Alpenstraße am 6. September 1934), *Die Straße*, no. 1 (September 1934): 88–91; Skåden, "The Map and the Territory." For beer and cigars, see the request for reimbursement. Ministerpräsident Sieber to Fritz Todt, September 25, 1934, Bayerisches Hauptstaatsarchiv StK 6948.

111. Amtliches Bayerisches Reisebüro, *75 Jahre abr*, 48; Seely, "Der amerikanische Blick," 19.

112. Simonson and Royall, *Landschaftsgestaltung*; Fritz Todt an die wegebaupflichtigen Länder und preussischen Provinzen, August 31, 1935, Staatsarchiv Augsburg, Straßen- und Flußbauamt Kempten, 106.

113. Todt an Ministerium des Innern, Ministerialabteilung, Munich, August 22, 1936, Staatsarchiv Augsburg, Straßen- und Flußbauamt Kempten, 106.

114. "Bericht des Regierungsbaurats Poppel, Referent für die Deutsche Alpenstraße beim Straßen- und Flußbauamt Kempten," May 31, 1937, Staatsarchiv Augsburg, Straßen- und Flußbauamt Kempten, 106.

115. Todt to von Keudell, February 3, 1937, Bundesarchiv R 4601/412, 3; Akbal, "'Trinkt, O Augen, was die Wimper hält.'" It appears that Todt had read two papers by one of his engineers on American roads who claimed that scenic driving had all but replaced hiking in the United States. Wehner, "Straße und Landschaft in den Vereinigten Staaten"; Wehner, "Parkstraßen in den Vereinigten Staaten". In addition to the Alpine Road and the Black Forest Road, at least two more scenic roads were built under Nazi rule: one in the Bavarian part of the Rhön Mountains and one in the mountains of Silesia. Prisoners of war and forced

laborers worked on the latter. Mölter, *Hochrhönstraße*; Max Veit, "Die deutsche Sudeten-straße," *Die Straße*, no. 16 (1936): 514–18; Wiater, "Droga Sudecka (Sudetenstraße.)"

116. Seidler, *Fritz Todt*, 19. For a helpful summary of a complex debate among histori-ans regarding the particularist Nazi ideology of "German Technology" (*Deutsche Technik*), see Guse, "Nazi Technical Thought Revisited" and the literature cited in this paper.

117. Brüggemeier, Cioc, and Zeller, *How Green Were the Nazis?*

118. Alwin Seifert, "Gebirgsstraßen der Schweiz," 11, January 27, 1936, Bundesarchiv Berlin R4601/142.

119. See the correspondence and memoranda in Bundesarchiv Berlin R4601/400 and 404.

120. "Hitler Highway Best in Europe," *Pittsburgh Sun-Telegraph*, February 16, 1935, clip-ping in Bundesarchiv Berlin, R4601/1506. It should be noted, however, that this newspaper article did not deal with this particular controversy. Alwin Seifert confirmed in a 1940 letter that the efforts for roadside protection had failed: Seifert to Staatliche Bauleitung der Deutschen Alpenstraße, Baustelle Resten, Obersalzberg, January 10, 1940, Alwin Seifert papers 223, Technical University Munich-Weihenstephan, Chair for Landscape Architecture and Public Space.

121. Pierce, "Road to Nowhere," 197–99.

122. Starnes, *Creating the Land of the Sky*, ch. 2: "Building Image and Infrastructure."

123. Silver, *Mount Mitchell*, 173–80. Mt. Washington in New Hampshire has been acces-sible by road since 1861: https://mtwashingtonautoroad.com/history, accessed July 23, 2018.

124. Whisnant, *Super-Scenic Motorway*; Starnes, *Creating the Land of the Sky*.

125. Preston, *Dirt Roads to Dixie*; Whisnant, *Super-Scenic Motorway*.

126. Whisnant, *Super-Scenic Motorway*, 20–21.

127. Whisnant, *Super-Scenic Motorway*, 33–39; Reich, "Re-creating the Wilderness."

128. Whisnant, *Super-Scenic Motorway*, 38; Caro, *Power Broker*, 288–91, 426; LaFrank, "Real and Ideal Landscapes." Virginia and North Carolina had to pay for the right-of-way, while the federal government would be responsible for the construction and maintenance of the road.

129. I am indebted to John Beardsley (Dumbarton Oaks) for helping me clarify some of these issues. McClelland, *Building the National Parks*.

130. Davis, *National Park Roads*.

131. Whisnant, *Super-Scenic Motorway*, ch. 2.

132. Raymond Loughlin, New York, NY, Green Mountain Club, to Secretary Ickes, Feb-ruary 11, 1934, NARA RG 79, Central Classified Files 2713. On MacKaye, see Sutter, *Driven Wild*, 142; Anderson, *Benton MacKaye*; Mittlefehldt, *Tangled Roots*.

133. MacKaye and Mumford, "Townless Highways." MacKaye had published a con-densed version by himself a year earlier: Benton MacKaye, "The Townless Highway," *The New Republic*, March 12, 1930.

134. Gas locomotives: quoted after Kates, "Path Made of Words," 128. Anderson, *Mac-Kaye*, 218–19; Sutter, *Driven Wild*, 142–93; Sutter, "'A Retreat from Profit.'" MacKaye's em-phasis on safety reinforces Sutter's claims regarding MacKaye's concerns for "labor and community stability" (*Driven Wild*, 142) and should be seen against the background of the animated discussion of automotive fatalities in the interwar period. Norton, *Fighting Traffic*.

135. Sutter, *Driven Wild*, 143, 162; Mumford, *Technics and Civilization*, 237–9. For important essays on Mumford, see Hughes and Hughes, *Lewis Mumford*. On electricity, see Nye, *Electrifying America*. For a critique of Mumford's thinking, see White, *Organic Machine*, 63–68.

136. Anderson, *MacKaye*, 232.

137. MacKaye, "Flankline vs. Skyline"; Sutter, *Driven Wild*; Anderson, *MacKaye*.

138. Sutter, *Driven Wild*, 179. The historical and historiographical discussion of wilderness is extensive. For examples, see Nash, *Wilderness and the American Mind*; Cronon, "Trouble with Wilderness"; Lewis, *American Wilderness*.

139. Whisnant, *Super-Scenic Motorway*, ch. 2; Whisnant, "The Scenic Is Political."

140. Whisnant, *Super-Scenic Motorway*, 68; R. Getty Browning to Theodore Strauss, Regional Advisory Board, January 17, 1934 (copy), NARA RG79, Central Classified Files, Box 2711; North Carolina Committee on Federal Parkway, *Description of a Route through North Carolina as a Part of the Scenic Parkway to Connect the Shenandoah National Park with the Great Smoky Mountains National Park*, 1934, n.p., NARA RG79, Central Classified Files, Box 2711.

141. Whisnant, *Super-Scenic Motorway*, 71, 98.

142. Historic American Engineering Record, *Blue Ridge Parkway* HAER no. NC-42, 1997, p. 40, 45, available at http://lcweb2.loc.gov/master/pnp/habshaer/nc/nc0400/nc0478/data /nc0478data.pdf, accessed October 8, 2019. The name "Blue Ridge Parkway" was chosen later, but I use it here for reasons of simplicity. Whisnant, *Super-Scenic Motorway*. Apparently, Clarke's work with the National Park Service as an advisor ended abruptly over a pay dispute. Oral History Interview of Stanley W. Abbott by S. Herbert Evison, 1958, Blue Ridge Parkway Archives, Asheville, NC, 9–10.

143. Thomas C. Vint, Chief Architect, to the Director, National Park Service, June 8, 1934, NARA RG 79, Central Classified Files 2711.

144. Oral History Interview of Stanley W. Abbott by S. Herbert Evison, 1958, Blue Ridge Parkway Archives, Asheville, NC, 12.

145. "Resident Landscape Architect, Report: Proposed Locations Shenandoah-Great Smoky Mountains National Parkway, to Chief Landscape Architect," n.p., National Park Service, June 8, 1934, NARA RG 79, Central Classified Files 2711. In private correspondence, MacKaye derided Abbott's group as "landscape artichokes" content with "planting a few pansies." Sutter, *Driven Wild*, 191. While most voices against higher altitudes cited scenic or wilderness values, one landscape architect and park promoter, Harlan Kelsey, had spoken of "the ecological value of the Park" when opposing the location of the Skyline Drive. Kelsey to MacKaye, March 27, 1925, cited after Dalbey, *Regional Visionaries*, 118–19. On Kelsey, see Whisnant, *Super-Scenic Motorway*, 280–82.

146. E. G. Frizzell, President, Smoky Mountains Hiking Club, Knoxville, TN, to Strauss, April 12, 1934, NARA RG 79, Central Classified Files 2711.

147. They were the crest of the Blue Ridge between Shenandoah and the Peaks of Otter in Virginia and the Pisgah Ledge and Balsam Mountains in North Carolina. Historic American Engineering Record, *Blue Ridge Parkway*, 46–47; Sutter, *Driven Wild*, 231–33; Davis, *National Park Roads*, 197.

148. Historic American Engineering Record, *Blue Ridge Parkway*, 49; Whisnant, *Super-Scenic Motorway*; Anderson, *MacKaye*; Maher, *Nature's New Deal*. After ten years of lobby-

ing, Tennessee convinced Congress to fund the project of a Foothills Parkway, a loop road around the northern end of Great Smoky Mountains National Park, but it is yet unfinished. Historic American Engineering Record, "Great Smoky Mountains National Park," 81–85; Pierce, "Road to Nowhere." MacKaye tried his hand at translating his planning ideas into reality at the TVA. Brian Black, "Organic Planning."

149. Historic American Engineering Record, *Blue Ridge Parkway*, 51–53, 63, 67; Oral History Interview of Stanley W. Abbott, 3–4, 13–14.

150. Wilson, *Culture of Nature*, 36.

151. Historic American Engineering Record, *Blue Ridge Parkway*, 7–8. Unsuccessfully, Abbott had lobbied for including Virginia's Natural Bridge rather than the Blue Ridge mountains in the Virginia routing. Historic American Engineering Record, Blue Ridge Parkway, 55–56; Oral History Interview of Stanley W. Abbott, 13. In North Carolina, R. Getty Browning, the state highway engineer, was more influential for the routing than Abbott, according to Whisnant, *Super-Scenic Motorway*.

152. Sutter, *Driven Wild*, 232–33; Pierce, "Road to Nowhere"; Historic American Engineering Record, Great Smoky Mountains National Park, 23; Marshall, "Memorandum for the Secretary," June 9, 1935; Ickes to Demaray, June 14, 1935 (copy), both NARA RG79, Central Classified Files, Box 2714. The letter by Ickes instructs Park Service associate director Arthur Demaray to consider Marshall's recommendations and consult with Ickes if they were not heeded. Confusingly, Ickes's letter refers to the Blue Ridge Parkway, but Marshall's memorandum deals with the Great Smoky Mountains National Park. By the time the memo and letter were written, the foundational routing decisions on the former had already been made.

153. Wilson, *Culture of Nature*, 35.

154. Reich, "Re-Creating the Wilderness," puts the number of displaced Virginians at 5,000; a website maintained by an activist group mentions 3,000. www.blueridgeheritage project.com, accessed April 23, 2021. See also Wilson, *Culture of Nature*, 35; Historic American Engineering Record, *Blue Ridge Parkway*, 63; Pierce, *Great Smokies*; Eisenfeld, *Shenandoah*. For other instances of displacing locals during the creation of state and national parks, see Jacoby, *Crimes Against Nature*; Spence, *Dispossessing the Wilderness*.

155. Whisnant, *Super-Scenic Motorway*, 242–43, 255–56. Davis correctly notes that the combination of scenery and living history was novel at the time. Davis, *National Park Roads*, 195, 199.

156. Oral History Interview of Stanley W. Abbott, 22; Whisnant, *Super-Scenic Motorway*, details the controversies over land purchases. In addition to purchasing land, planners entered into scenic easements with owners of adjacent lands to regulate land use for view protection. Whisnant, *Super-Scenic Motorway*, 117; Oral History Interview of Stanley W. Abbott, 29–30; Davis, "Protecting Scenic Views."

157. Whisnant, *Super-Scenic Motorway*; Daniels, "Marxism, Culture, and the Duplicity of Landscape." On prohibiting ambulances, see NPS Regional Director, Richmond, VA, to Superintendent, Natchez Trace Parkway (copy), December 31, 1953, Blue Ridge Parkway Archives, Asheville, North Carolina, RG11 Records of the Interpretive Specialist BLRI-440 CAT19859, Series 3, Box 36, Folder 8.

158. Oral History Interview of Stanley W. Abbott, 24, 41–43; Whisnant, *Super-Scenic*

Motorway, 151–52; Maher, *Nature's New Deal*; Gregg, *Managing the Mountains*; Phillips, *This Land, This Nation*. On the precarious role of wildlife biologists and ecologists in the National Park Service during the New Deal, see Sellars, *Preserving Nature*, 147.

159. George Keeley, "Blue Ridge Parkway Adds Vital Link," *New York Times*, June 5, 1960, XX18.

160. Cited after Davis, *National Park Roads*, 181. For an international perspective on New Deal politics, see Patel, *New Deal*.

161. Wohl, *Passion for Wings*; Asendorf, *Super Constellation*; Dümpelmann, *Flights of Imagination*.

162. In a letter to Benton MacKaye, National Park Service director Cammerer reminded him that "for every one or two of you [avid, physically fit hikers] there are perhaps a thousand who cannot take these hikes and to whom some opportunity must be afforded to see the beautiful scenes within the park." Cammerer to MacKaye, September 14, 1934, NARA RG79, Central Classified Files, Box 2714.

163. Rosalie Edge, *Roads and More Roads in the National Parks and National Forests* (New York: Emergency Conservation Committee, 1936), 4–5, NARA RG 79, Central Classified Files 448; Schwenkel, "Motorisierung und Naturschutz."

164. Gassert, *Amerika im Dritten Reich*; de Grazia, *Irresistible Empire*; Nolan, *Transatlantic Century*.

Chapter 4. Roads out of Place

1. Paul J. C. Friedlander, "The Traveler's World: Pax Americana in the Blue Ridge," *New York Times*, September 12, 1971, XX31.

2. A. E. Demaray, Acting Director, NPS, and Thomas H. MacDonald, Chief, BPR, Memorandum for the Secretary, May 23, 1935, NARA RG 79, Central Classified Files 448; LaFrank, "Real and Ideal Landscapes," 254–55; Gandy, *Concrete and Clay*, 122; James W. Curley, Governor of Massachusetts, to Ickes, May 24, 1935; Ickes to Curley (copy), June 7, 1935, both NARA RG 79, Central Classified Files 448.

3. Benton J. Stong, "Highway of 20,000 Miles to Link National Parks is Proposed," *New York World Telegram*, April 4, 1935, clipping in NARA RG 79, Central Classified Files 448; "An Adequate Highway Program Including Provision for a National Parkway Proposal. Extension of Remarks of Hon. Wilbur Cartwright of Oklahoma in the House of Representatives," June 16, 1934, NARA RG 79, Central Classified Files 448.

4. The lone landscape architect was Charles W. Eliot II, who at the time worked for the National Resources Planning Board in Washington, DC. "Proposed National 'Tourways' Plan," "Proposed Personnel of National Tourways Committee of the Construction League of the United States," n.d. [probably July 1934], both in NARA RG 79, Central Classified Files 448. On Eliot, see Davis, *National Park Roads*, 88.

5. Dilsaver, *America's National Park System*, 136.

6. Roland Rice, "Roads Were Built for Commerce not Sightseeing," *Commercial Car Journal with which is Combined Operation & Maintenance* 56 (November 1938): 118; Fruehauf Trailer Co., *Public Servant No. 1: 10 Challenging Statements That Tell the Truth about Highways* (Detroit, MI: Fruehauf Trailer Co., 1949), 9, Hagley Museum and Library, Wilmington, Delaware; Rose and Mohl, *Interstate*, 9; Hamilton, *Trucking Country*.

7. Rose and Mohl, *Interstate*, 9–11; Seely, *Building the American Highway System*; Rose, Seely, and Barrett, *Best Transportation System*.

8. Public Roads Administration, "Toll Roads and Free Roads," 11, 39, 52, 91, 96.

9. Corn, Horrigan, and Chambers, *Yesterday's Tomorrows*; Marchand, "The Designers Go to the Fair II"; Nye, *American Technological Sublime*, 218–20; Seely, *Building the American Highway System*, 170; Speck, "Futurama."

10. Max-Erich Feuchtinger, "Ein alpenländisches Autowanderstraßennetz," *Der Straßenbau* 28, no. 17 (September 1, 1937): 215–20, 231–35; Karl Gustav Kaftan, "Von der Reichsautobahn zur Europa-Autobahn," *Der Straßenbau* 28, no. 4, 42–47. On European road networks, see Schipper, *Driving Europe*.

11. Klenke, *Bundesdeutsche Verkehrspolitik*; Möser, *Geschichte des Autos*.

12. Seiler, *Republic of Drivers*.

13. Zeller, "Loving Parks, Embracing Highways."

14. Seely, " 'Push' and 'Pull' Factors"; Oreskes and Krige, *Science and Technology*; Krige, *How Knowledge Moves*.

15. Kemp, "Aesthetes and Engineers"; Seely, *Building the American Highway System*; Lewis, *Divided Highways*; Swift, *Big Roads*; Petroski, *The Road Taken*.

16. New Jersey Turnpike, *Garden State Parkway*. In 2007, environmentalists protested the widening of this road. Robert Strauss, "Parkway Expansion Plan Draws Opposition," *New York Times*, May 13, 2007.

17. Gillespie and Rockland, *Looking for America*.

18. Davis, *National Park Roads*, 201–6, 209–12.

19. Dalbey, *Regional Visionaries*, ch. 6, 139–76; Davis, *National Park Roads*, 212–15.

20. Department of Commerce, "Parkway for the Mississippi"; Abbott, "Parkways—A New Philosophy"; Great River Road, "Welcome to the Great River Road."

21. Achenbach, *Grand Idea*.

22. Achenbach, *Grand Idea*, 283–84; Mackintosh, "Shootout on the Old C. & O. Canal"; Sowards, *Environmental Justice*. Davis, *National Park Roads*, 215–19, also stresses the significance of this defeat.

23. United States Department of Commerce, "A Proposed Program," 140; United States Federal Highway Administration, "An Assessment of the Feasibility"; Zimring, " 'Neon, Junk, and Ruined Landscape' "; Gudis, *Buyways*; US Department of Transportation, "Scenic Byways"; Federal Highway Administration, *America's Byways*. The practice of naming existing roads to attract tourists remains widespread in Germany, where dozens of named roads exist, from the Romantic Road in Southern Germany to the Asparagus Road and the Beer and Castles Road. The Romantic Road was inaugurated in 1950 and has been very successful, especially with American and Japanese tourists. Strache, *Romantische Strasse*. For a listing of theme roads, see "Das Ferienstraßennetz."

24. Rugh, *Are We There Yet?*; Davis, *National Park Roads*, 222–43; Carr, *Mission 66*; McClelland, *Building the National Parks*; Catton, "Hegemony"; Historic American Engineering Record, Great Smoky Mountains National Park Roads and Bridges, 7.

25. Edward Abbey, *Desert Solitaire: A Season in the Wilderness* (New York: McGraw-Hill, 1968), 59.

26. For treatments of national park policy, see Miles, *Wilderness in National Parks*;

Havlick, *No Place Distant*. A nuanced case study of Abbey and roadbuilding can be found in Rogers, *Roads in the Wilderness*. For wilderness policies in general, see Turner, *Promise of Wilderness*.

27. Schultz, "To Render Inaccessible." It is worth noting that local boosters proposed a high-altitude Sierra Way in the 1920s and 1930s, with a length of more than five hundred miles and altitudes of six thousand feet or more for most of the route. A study by the BPR concluded that it was possible to build the road, but that it would be very expensive. The NPS opposed the road, and it was never built. Dilsaver and Tweed, *Challenge of the Big Trees*, ch. 6.

28. Whisnant, *Super-Scenic Motorway*, 263–325.

29. Dilsaver, *America's National Park System*, 147–48. By 1942, national parks in the South had to follow the Department of the Interior's policy on the "non-discrimination of Negroes."

30. Young, "'A Contradiction'"; O'Brien, *Landscapes of Exclusion*; Finney, *Black Faces, White Spaces*; Taylor, *Overground Railroad*, 167; Kahrl, *The Land Was Ours*.

31. Sorin, *Driving While Black*; Seiler, "So That We."

32. Jones, *Historic Resource Study: African Americans*, 13; Walker, "DeHaven Hinkson."

33. Kania-Schütz, "Einleitung," 9.

34. "Der Fremdenverkehr—das A und O der Berchtesgadener Wirtschaft. Stellungnahme des Fremdenverkehrsausschusses des Gemeinderats Berchtesgaden zum Fremdenverkehr," February 11, 1946; Staatsarchiv Munich LRA 198857.

35. "An der Alpenstraße wird wieder gebaut," *Neue Zeitung*, June 27, 1953.

36. "Bayerischer Landtag, Ausschuß für Wirtschaft und Verkehr, 46. Sitzung," June 13, 1952; "Alphabetisches Sach- und Sprechverzeichnis zu den Verhandlungen des Bayerischen Landtags in der Tagung 1951/52," both Archives of the Bayerischer Landtag.

37. "Zäher Kampf um Kilometer," *Süddeutsche Zeitung*, October 18, 1951.

38. "An der Alpenstraße wird wieder gebaut," *Neue Zeitung*, June 27, 1953.

39. "Bayerischer Landtag, Ausschuß für Ernährung und Landwirtschafts, 123. Sitzung," September 9, 1954; "Alphabetisches Sach- und Sprechverzeichnis zu den Verhandlungen des Bayerischen Landtags in der Tagung 1953/54," both Archives of the Bayerischer Landtag. The committee did not reach a decision in the matter.

40. "Bayerischer Landtag, Ausschuß für Ernährung und Landwirtschaft, 5. Sitzung," February 22, 1955; "Alphabetisches Sach- und Sprechverzeichnis zu den Verhandlungen des Bayerischen Landtags in der 3. Wahlperiode 1954/58," both Archives of the Bayerischer Landtag. This time, the committee recommended that the state of Bavaria should build and maintain "simple" fences. The debate was concerned with Upper Bavaria only.

41. "Bayerischer Landtag, 5. Legislaturperiode, Beilage 2813, Schriftliche Anfrage Wilhelm Röhrl," and answer by Junker, home secretary, Archives of the Bayerischer Landtag.

42. Gelberg, "Oberste Baubehörde zwischen 1932 und 1949," 338–9; Gall, *Gute Straßen*.

43. Dieter Schröder, "Respekt vor dem, der Autobahnen baut," *Der Spiegel*, no. 46, 1964, 47–53, 47. http://wissen.spiegel.de/wissen/image/show.html?did=46176040&aref=image036 /2006/03/06/cqsp196446047-P2P-053.pdf&thumb=false, accessed July 23, 2012. For Seebohm's role as a revisionist spokesperson for Sudeten Germans, see Ahonen, *After the Expulsion*.

44. Gerhard Tomkowitz, "Seebohm öffnet ein Stück Alpenstraße," *Süddeutsche Zeitung*, July 17, 1959. The Alpine Road even made it into a 1957 federal plan for roadbuilding, but as a tourist road, not a regular road. Bundesminister, *Ausbauplan*, Attachment 13. The plan envisioned 90 miles (145.1 kilometers) of new construction for the Alpine Road and 68 miles (110 kilometers) for the Black Forest High Road. As the historian Alexander Gall argues, the plan had the character of a wish list (with requests supplied by individual states) rather than indicating a commitment by the federal government to plan, finance, and build the highways. Gall, *Gute Straßen*, 143–45. "Niederschrift über die Besprechung am 23. Juni 1965 um 19 Uhr im Sitzungssaal der Obersten Baubehörde in München mit dem Herrn Bundesverkehrsminister Dr.-Ing. Hans-Christoph Seebohm und Vertretern der Interessengemeinschaft Deutsche Alpenstraße," Bayerisches Hauptstaatsarchiv, Bevollmächtigter Bayerns beim Bund 977. Seebohm's referral to 134 kilometers of road was somewhat disingenious, since it included a stretch of new autobahn to Austria via the Inn valley and two bypass roads around cities. For the lobbying group, see Bayerisches Wirtschaftsarchiv Munich, K 9/732, "Industrie und Handelskammer Schwaben (Augsburg)," Folder "Interessengemeinschaft Deutsche Alpenstraße." Austria did not join the European Union until 1995.

45. Schürmann to Leber, January 1, 1967, Bundesarchiv Koblenz B108/62301; Semmens, *Seeing Hitler's Germany*; Spode, "Fordism, Mass Tourism"; Baranowski, *Strength through Joy*; Confino, "Traveling as a Culture"; Kopper, "Breakthrough of the Package Tour"; Denning, *Skiing into Modernity*.

46. "Ein Stück Deutsche Alpenstraße einmal nicht im Tal," *Nürnberger Zeitung*, February 9, 1965; "Niederschrift über die Besprechung am 23. Juni 1965", 6. Oberammergau's mayor Lang, as it turns out, was the same in 1965 as during the Nazi years. Having survived his denazification proceedings with a slap on the wrist, the town's voters in 1948 entrusted him with the same office he had under Hitler's reign. Utschneider, *Oberammergau*, 256. Waddy, *Oberammergau*, 245–7, cites evidence that Lang was a "mild" mayor during the Nazi years and trusted for his competence.

47. Deutscher Alpenverein to Seebohm, March 19, 1965, Bundesarchiv Koblenz B108/62301.

48. Verein zum Schutze der Alpenpflanzen und -tiere to Seebohm, March 15, 1965, Bundesarchiv Koblenz B108/62301.

49. On protests by Bavarian conservationists against what they called the "mountain-train plague," see Hasenöhrl, *Zivilgesellschaft*, 165–87. On conservation in postwar Germany, see Chaney, *Nature of the Miracle Years*; Engels, *Naturpolitik*.

50. Karl-Heinz Kaschel, "Alpenstraße sollte über Hochgrat führen," *Allgäuer Zeitung*, September 17, 1988.

51. Lekan and Zeller, "Region, Scenery, and Power"; Sutter, *Driven Wild*.

52. Bernhard Magass, "Deutsche Alpenstraße eine ewige Baustelle," *Allgäuer Zeitung*, September 9, 1978.

53. Armin Ganser, "Jubiläumstrip über eine 'schöne Unvollendete,'" *Süddeutsche Zeitung*, August 9, 1988. In a case of historical amnesia, the monthly magazine *Bayerland* (founded in 1890) claimed that the road lay fallow from 1927 to 1950: Günter D. Roth, "Deutsche Alpenstraße, Teil Ost: Von Berchtesgaden bis Mittenwald," *Bayerland* 71, April 1969,

2–15, 7; Günter D. Roth, "Deutsche Alpenstraße, Teil West: Von Garmisch-Partenkirchen bis Lindau," *Bayerland* 71, May 1969, 2–17.

54. Stefan König, "'Perlenfischer' geben sich zuletzt die Kugel," *Münchner Merkur*, May 18, 2002; "Die NS-Zeit in Wolfratshausen," http://www.braun-in-wolfratshausen.de/5.html, accessed July 24, 2012; Karl Stankiewitz, "450 Kilometer Panorama," *Süddeutsche Zeitung*, May 16, 2002, 51.

55. Koshar, "'What Ought to Be Seen'"; Jakle, *The Tourist*; Pagenstecher, *Der bundesdeutsche Tourismus*.

56. Schmithals, *Die Deutsche Alpenstraße. Vom Bodensee zum Königssee*. Seebohm's preface is on p. 5. Echoing Todt's parlance, he calls a properly landscaped road a "cultural achievement."

57. Mair, *Hochstraßen*, 57–80.

58. Müller-Alfeld, *Die Deutsche Alpenstrasse*, 129.

59. Müller-Alfeld, *Die Deutsche Alpenstrasse*, 16.

60. The first image of the road is on p. 22 out of 135.

61. Strache, *Autoparadies*, 3.

62. Prager, *Deutsche Alpenstrasse*.

63. Mittermeier and Hirschbichler, *Traumlandschaften*.

64. My analysis is confined to the guidebooks on the Blue Ridge Parkway available at the Library of Congress as of 2007.

65. National Park Service, *Blue Ridge Parkway*; Lord, *Blue Ridge Parkway Guide* (Asheville, NC: 1959–63); Lord, *Blue Ridge Parkway Guide* rev. ed. (New York: 1982).

66. Catlin, *Naturalist's Blue Ridge Parkway*, 1; Humphries, *Along the Blue Ridge Parkway*; Rives, *Blue Ridge Parkway*; Humphries, *Images*.

67. Shelton-Roberts and Roberts, *Blue Ridge Parkway*.

68. Chilcoat, *Lonely Plant Road Trip*, 4.

69. Schrag, "The Freeway Fight," 650; Caro, *Power Broker*; Lewis, *Divided Highways*; Rose and Mohl, *Interstate*; Rose and Seely, "Getting the Interstate System"; Mohl, "Stop the Road"; Ballon and Jackson, *Robert Moses*; Gutfreund, *Twentieth Century Sprawl*; Gioielli, *Environmental Activism*; Avila, *Folklore of the Freeway*.

70. Mumford, *Myth of the Machine*, caption to Panel 22, "Organized Destruction," between pages 340 and 341.

71. Ladd, *Autophobia*, 113–17.

72. Davis, "Rise and Decline"; Grady Clay, "The Tiger Is Through the Gate," *Landscape Architecture* 49 (Autumn 1958): 79–82.

73. Halprin, *Life Spent Changing Places*. I am indebted to John Beardsley for sharing his views of Halprin with me.

74. Cultural Landscape Foundation, *Lawrence Halprin Oral History Interview Transcript*, 79. The interview was conducted in 2003.

75. For details of Repton's designs, see Daniels, *Humphry Repton*.

76. Halprin, *Freeways*, 29. For context, see Tunnard and Pushkarev, *Man-Made America*; Snow, *Highway and the Landscape*.

77. Halprin, *Freeways*, 12, 17 ("Freeways which carry the automobile in its adventures are amongst the beautiful structures of our age"), 27.

78. Halprin, *Freeways*, 113.

79. "To Highway Consultants Group," Halprin, *Notebooks*, 156–57. The entry is not dated, but it appears before one dated March 1966. Twenty-four years prior, Gilmore Clarke had assigned blame for the powerlessness of landscape architects vis-à-vis engineers on the former, since they were not "sufficiently well trained to assume the degree of responsibility in such an enterprise such as the Pennsylvania Turnpike." Clarke, "Beauty, a Wanting Factor."

80. Merriman, "Roads: Lawrence Halprin."

81. Jackson, "Abolish the Highways!"; Jackson, "Abstract World"; Davis, "Looking Down the Road"; Venturi, Brown, and Izenour, *Learning from Las Vegas*. For an intellectual history of changing views of the American roadside, see Esperdy, *American Autopia*.

82. Urban Advisors, "Freeway in the City"; Annese, *Office of Clarke and Rapuano*; Mary Perot Nichols, "Private Opinion: Wanted: A Philosophy of Transportation," *Village Voice*, July 24, 1969. Clarke and Rapuano had contributed to designing the Major Deegan Expressway, the Brooklyn-Queens Expressway, and the Van Wyck Expressway, which were commercial roads in dense neighborhoods. The *Village Voice* article was one of many critical voices against the planned Lower Manhattan Expressway. Caro, *Power Broker*.

83. Rome, *Genius of Earth Day*, 117–20, 118 (quote).

84. Commoner, "Technology and the Natural Environment."

85. Wilson correctly calls the Blue Ridge Parkway a "prototypical environment of instruction." Wilson, *Culture of Nature*, 35. Lynn Nisbet, "Side Roads Offer Scenic Views," *The Daily Times* (Burlington, NC), July 30, 1959, 3.

86. Joseph C. Ingraham, "White-on-Green Highway Signs Chosen by Autoists in U.S. Poll," *New York Times*, January 27, 1958; Joshua Yaffa, "The Road to Clarity," *New York Times Magazine*, August 12, 2007, available at http://www.nytimes.com/2007/08/12/magazine/12 fonts-t.html?_r=0&pagewanted=print, accessed October 18, 2019; Vanderbilt, *Traffic*, 90.

87. For some examples of the literature on automotive fatalities and safety, see Blanke, *Hell on Wheels*; Norton, *Fighting Traffic*; Vinsel, *Moving Violations*; Zeller, "Mein Feind, der Baum"; Zeller, "Loving the Automobile to Death."

88. Mom, *Atlantic Automobilism*; Bijsterveld et al., *Sound and Safe*; Rugh, *Are We There Yet?*; Fabian, *Boom in der Krise*, 225–97. Of course, accounts on the aesthetic appreciation of driving are still published. For a sensitive journey by two Brutalist architects, see Smithson, *AS in DS*.

89. For two examples of this genre, see Pirsig, *Zen and the Art*; Heat Moon, *Blue Highways*.

Epilogue

1. Kroll, "Environmental History of Roadkill"; Swart, "Reviving Roadkill."

2. For a broader history, see Anker, *From Bauhaus to Ecohouse*.

3. For an envirotechnical analysis of large-scale projects and displacement, see Parr, *Sensing Changes*.

4. Thomas P. Hughes speaks of "technological enthusiasm." Hughes, *American Genesis*. For more extensive visions, see Segal, *Technological Utopianism*.

5. Tarr, *Search for the Ultimate Sink*, ch. 10, "The Horse: Polluter of the City," 323–34; McShane and Tarr, *Horse in the City*; Greene, *Horses at Work*.

6. Robert Marshall, "Memorandum for the Secretary," June 14, 1935, NARA RG 79, Central Classified Files, Box 448.

7. For the baby boom generation, Jack Kerouac was the touchstone of this attitude. Kerouac, *On the Road*. Some examples of the vast literature on road movies include Laderman, *Driving Visions*; Sargeant and Watson, *Lost Highways*.

8. Garraty, "New Deal, National Socialism"; Schivelbusch, *Three New Deals*.

9. R. Getty Browning, Senior Locating and Claim Engineer, to A. E. Demaray, Associate Director, NPS, March 14, 1938; Demaray to Browning, March 24, 1938, Browning to Demaray, April 8, 1938, all NARA RG 79, Central Classified Files, Box 2715. Browning relayed local perceptions that the parkway would be exclusive and for outsiders. Demaray responded that a newsletter to residents would assuage these concerns.

10. "Provisions for Negroes," Annual Report of the Blue Ridge Parkway to the Director, NPS, June 30, 1940, 17, Blue Ridge Parkway Archives, Asheville, NC; Dilsaver, *America's National Park System*, 143–48; O'Brien, *Landscapes of Exclusion*.

11. Hillory A. Tolson, Assistant Director, Memorandum for Mr. Demaray, August 25, 1944, NARA RG 79, Central Classified Files, Box 449.

12. Sutter, *Driven Wild*.

13. Roth and Divall, *From Rail to Road and Back Again?*

14. Ron Shaffer, "Scenery or DVDs? Parents Take Sides," *Washington Post*, April 24, 2005, C2.

15. Cresswell, "Towards a Politics," 23. Even though the author discusses the slow food movement, the question appears pertinent here.

16. Kern, *Culture of Time and Space*; Borscheid, *Tempo-Virus*; Kaschuba, *Überwindung der Distanz*; Benesch, "Dynamic Sublime"; Rosa, "Social Acceleration"; Ogle, *Global Transformation of Time*.

17. Schivelbusch, *Railway Journey*, 33–44.

Bibliography

Primary Sources

Archives: United States
Asheville, North Carolina, Blue Ridge Parkway Headquarters
    Blue Ridge Parkway Archives
College Park, Maryland, National Archives II
    National Park Service (RG79)
        Central Classified File
        Blue Ridge Parkway
    Bureau of Public Roads (RG30)
        Text Files
        Image Files
    Berlin Document Center
Fairfax, Virginia, William L. Mertz Transportation Collection, Collection #C0050, Special
    Collections and Archives, George Mason University Libraries
New York, Columbia University, Reminiscences of Gilmore D. Clarke (1959), Oral History
    Archives at Columbia, Rare Book & Manuscript Library
Philadelphia, Free Library, Automobile Reference Collection
Philadelphia, University of Pennsylvania, Lewis Mumford Papers
Washington, DC, Library of Congress, Papers of Woodrow Wilson
Wilmington, Delaware, Hagley Library and Collections

Archives: Germany
Augsburg, Staatsarchiv (State Archive)
    Bezirksamt Füssen (District Office Füssen)
Berlin, Bundesarchiv (Federal Archives)
    Generalinspektor für das deutsche Straßenwesen (General Inspector for German Roads)
Freising-Weihenstephan, Technical University of Munich, Lehrstuhl für Landschaftsarchi-
    tektur und öffentlichen Raum (Chair for Landscape Architecture and Public Space)
    Alwin Seifert Papers
Füssen, Stadtarchiv (City Archive)
    Files on German Alpine Road
Garmisch-Partenkirchen, Marktarchiv (City Archive)
    Files on German Alpine Road
Inzell, Gemeindearchiv (Town Archive)
    Files on German Alpine Road

Kempten, Staatliches Bauamt (State Construction Office)
  Files on German Alpine Road
Koblenz, Bundesarchiv (Federal Archives)
  Reichsstelle für Naturschutz B245 (Reich Office for Conservation)
  Bundesminister für Verkehr B108 (Department of Transportation)
Munich, Bayerisches Hauptstaatsarchiv (Bavarian Main State Archives)
  Landtag (State Assembly)
  Staatskanzlei (Office of the Prime Minister)
  Staatsministerium für Handel, Industrie und Gewerbe (Department of Commerce, In-
    dustry, and Manufacturing)
Munich, Bayerischer Landtag (Bavarian State Assembly)
  Archives
Munich, Bayerisches Wirtschaftsarchiv
  Industrie und Handelskammer Schwaben (Augsburg), Folder Interessengemeinschaft
    Deutsche Alpenstraße
Munich, Deutscher Alpenverein (German Alpine Association)
  Archiv und Zeitschriften
Munich, Deutsches Museum, Archive
  Alwin Seifert Papers, NL 133
Munich, Staatsarchiv (State Archive)
  Bezirksämter Rosenheim, Traunstein, Berchtesgaden, Garmisch, Bad Tölz (District Of-
    fices Rosenheim, Traunstein, Berchtesgaden, Garmisch, Bad Tölz)
  Denazification file Josef Vilbig, Spruchkammern, Karton 1865
Munich, Stadtarchiv (City Archives)
  Fotosammlung (Photograph Collection)
Prien am Chiemsee, Marktarchiv (City Archive)
  Biography Dr. August Knorz
Ramsau bei Berchtesgaden, Gemeindearchiv (Town Archive)
  Files on German Alpine Road
Traunstein, Straßenbauamt (Road Building Office)
  Files on German Alpine Road

Printed and Electronic Sources

Journals
Better Roads
Landscape Architecture
Motor-Tourist
Die Straße (1934–42)

Newspapers
New York Times
Washington Post

Books, Journal Articles, and Internet Sources

Abbey, Edward. *Desert Solitaire: A Season in the Wilderness*. New York: McGraw-Hill, 1968.

Abbott, Stanley W. "Parkways—A New Philosophy." *American Planning and Civic Annual* (1951): 41–45.

Abraham, George D. *Motor Ways at Home and Abroad*. London: Methuen, 1928.

Amtliches Bayerisches Reisebüro. *75 Jahre abr. Rückschau und Ausblick*. Munich: abr, 1985.

Annese, Domenico. *The Office of Clarke and Rapuano, Inc*. New York: Clarke and Rapuano, 1972.

"Bayerischer Landtag: Verhandlungen 1919–1933." http://geschichte.digitale-sammlungen .de/landtag1919/band/bsb00008701.

Belloc, Hilaire. *The Road*. Manchester, UK: Charles W. Hobson, 1923.

Bierbaum, Otto Julius. *Eine empfindsame Reise im Automobil. Von Berlin nach Sorrent und zurück an den Rhein*. Munich: Wilhelm Goldmann, 1955 [orig. 1903].

British Road Federation. *Landscaping of Motorways. A Conference at the Institution of Civil Engineers*. London: British Road Federation, 1962.

Bundesminister für Verkehr. *Ausbauplan für die Bundesfernstrassen (Bundesstrassen und Bundesautobahnen)*. Bonn: Bundesminister für Verkehr, 1957.

Bureau of the Census. "Census Questionnaire Content, 1990 CQC-26." https://www.census .gov/prod/cen1990/cqc/cqc26.pdf, accessed July 20, 2017.

Bureau of Public Roads, Department of Commerce, and Department of the Interior National Park Service. *Parkway for the Mississippi*. Washington, DC: Government Printing Office, 1951.

Casagrande, Leo. *Bodenmechanik und neuzeitlicher Straßenbau*. Berlin: Volk und Reich, 1936.

Catlin, David T. *A Naturalist's Blue Ridge Parkway*. Knoxville: University of Tennessee Press, 1984.

Chilcoat, Loretta. *Lonely Planet Road Trip: Blue Ridge Parkway*. Oakland, CA: Lonely Planet Publications, 2005.

Christomannos, Theodor. *Die neue Dolomitenstrasse Bozen-Cortina-Toblach und ihre Nebenlinien*. Vienna: Reisser, 1909.

Clarke, Gilmore D. "Beauty, a Wanting Factor in the Turnpike Design." *Landscape Architecture* 32, no. 1 (January 1942): 53–54.

Commoner, Barry. "Technology and the Natural Environment" [1969]. In *Changing Attitudes toward American Technology*, edited by Thomas Parke Hughes, 54–62. New York: Harper and Row, 1975.

The Cultural Landscape Foundation, Pioneers of American Landscape Design. Lawrence Halprin Oral History Interview Transcript. http://72.27.230.88/sites/default/files/pioneers /halprin/videos/pdf/halprin_transcript.pdf, accessed May 11, 2010.

Davis, Chandler, and Gilmore C. Clarke. "Bridge-Building of the Sixth Engineers." *The Military Engineer* 17, no. 96 (November–December 1925): 487–90.

Dilsaver, Larry M. *America's National Park System: The Critical Documents*. 2nd ed. Lanham, MD: Rowman & Littlefield, 2016.

Dr. Todt: Berufung und Werk. Ein Film des Hauptamtes für Technik der NSDAP. Richard Scheinpflug, dir., 1942. http://www.youtube.com/watch?v=z4ZGZzzs5U8, accessed June 28, 2013.

Federal Highway Administration. "America's Byways." https://www.fhwa.dot.gov/byways, accessed May 25, 2021

"Das Ferienstraßennetz." https://www.ferienstrassen.info/, accessed July 23, 2018.

Feuchtwanger, Lion. *Erfolg. Drei Jahre Geschichte einer Provinz.* Berlin: Gustav Kiepenheuer, 1931.

Finger, F. A., and Waldemar Wucher, eds. *Die Alpenstraßen. Bauleistung und Verkehr in Geschichte und Gegenwart.* Berlin: Volk und Reich, 1937.

Fischer, Hans. *Bayern links und rechts der Alpenstraße.* Munich: Bergverlag Rudolf Rother, 1938.

Freeston, Charles L. *The Alps for the Motorist.* New York: Charles Scribner's Sons, 1927.

———. *The High-Roads of the Alps: A Motoring Guide to One Hundred Mountain Passes.* 2nd ed. New York: Charles Scribner's Sons, 1911. https://babel.hathitrust.org/cgi/pt?id=nyp.33433006689834&view=1up&seq=1, accessed October 1, 2019.

Giedion, Sigfried. *Space, Time and Architecture: The Growth of a New Tradition.* 5th ed. Cambridge, MA: Harvard University Press, 1967.

Great River Road. "Welcome to the Great River Road." https://experiencemississippiriver.com/, accessed July 27, 2018.

Gubbels, Jac L. *American Highways and Roadsides.* Boston: Houghton Mifflin, Riverside Press, 1938.

Halprin, Lawrence. *Freeways.* New York: Reinhold, 1966.

———. *A Life Spent Changing Places.* Philadelphia: University of Pennsylvania Press, 2011.

———. *Notebooks 1959–1971.* Cambridge, MA: MIT Press, 1972.

Heat Moon, William Least. *Blue Highways: A Journey into America.* Boston: Little, Brown, 1982.

Historic American Engineering Record, Blue Ridge Parkway. HAER no. NC-42, 1997. http://lcweb2.loc.gov/master/pnp/habshaer/nc/nc0400/nc0478/data/nc0478data.pdf, accessed October 8, 2019.

Historic American Engineering Record, Bronx River Parkway Reservation. HAER-NY 327, 2001. http://www.westchesterarchives.com/BRPR/Report_fr.html, accessed March 31, 2010.

Historical American Engineering Record, George Washington Memorial Parkway. HAER No. VA-69, 1998. http://lcweb2.loc.gov/pnp/habshaer/va/va1600/va1677/data/va1677data.pdf, accessed June 13, 2014.

Historic American Engineering Record, Great Smoky Mountains National Park Roads and Bridges. HAER TN-35, 1996.

Holitscher, Arthur. *Amerika heute und morgen. Reiserlebnisse.* 2nd ed. Berlin: S. Fischer, 1912. https://archive.org/stream/amerikaheuteundooholigoog#page/n9/mode/2up.

Humphries, George. *Along the Blue Ridge Parkway.* Englewood, CO: Westcliffe, 1997.

———. *Images of the Blue Ridge Parkway.* Greensboro, NC: Our State Books, 2005.

Ilf, Ilya, and Evgeny Petrov. *Ilf and Petrov's American Road Trip: The 1935 Travelogue of Two Soviet Writers.* Translated by Anne O. Fisher. New York: Cabinet Books/Princeton University Press, 2007.

International Road Congress. *Third International Road Congress, London 1913: Report of the Proceedings of the Congress.* Rennes-Paris: Oberthür, 1913.

Jackson, J. B. "Abolish the Highways!" *National Review* 29 (November 1966): 1213–1217.

———. "The Abstract World of the Hot-Rodder." In *Changing Rural Landscapes*, edited by Ervin H. Zube and Margaret J. Zube, 140–51. Amherst: University of Massachusetts Press, 1977.

James, H. M. "The Automobile and Recreation." *The Annals of the American Academy of Political and Social Science* 116, no. 1 (1924): 32–34.

Jones, Rebecca. *Historic Resource Study: African Americans and the Blue Ridge Parkway* (n.p.: National Park Service, 2009). http://www.npshistory.com/publications/blri/hrs-african -americans.pdf, accessed April 23, 2021.

Kerouac, Jack. *On the Road*. New York: Viking, 1957.

Kittel, Manfred. *Maximiliansweg. Auf der "Königsroute" von Lindau nach Berchtesgaden*. Munich: J. Berg, 1991.

Klemperer, Victor. *Ich will Zeugnis ablegen bis zum letzten. Tagebücher 1933–1941*. 5th ed. Berlin: Aufbau, 1996.

Knorz, [August] San.-Rat Dr. "Die bayerische Alpenstraße." *Nachrichtenblatt. Amtliches Organ des Gau Südbayern im Allgemeinen Deutschen Automobil-Club* 6, no. 3 (February 9, 1933); 3–6.

Knorz, [August] Sanitätsrat Dr. "Grundlagen des Fremdenverkehrs in Südbayern." *Völkischer Beobachter. Bayernausgabe*, September 11–12, 1932.

Lancaster, Samuel Christopher. *The Columbia: America's Great Highway through the Cascade Mountains to the Sea*. 3rd ed. Portland, OR: J. K. Gill, 1926.

"Der Landtag 1932–1933 (5. Wahlperiode)." https://www.bavariathek.bayern/medien-themen /portale/geschichte-des-bayerischen-parlaments/landtage-seit-1819.html, accessed January 5, 2022.

Lewis, Nelson P. Statement at the International Road Congress. Premier Congrès International de la Route, Compte rendu des travaux du congrès (Paris: Lahure, 1909), Session 1, 142–149, 147, available at https://babel.hathitrust.org/cgi/pt?id=mdp.39015074 988307;view=1up;seq=388.

Lord, William G. *Blue Ridge Parkway Guide*. 4 vols. Asheville, NC: Stephens Press, 1959–63.

———. *Blue Ridge Parkway Guide*. 4 vols., rev. ed. New York: Eastern Acorn Press, 1982.

Lynd, Robert S., and Helen Merrill Lynd. *Middletown: A Study in American Culture*. New York: Harcourt Brace, 1959.

MacKaye, Benton. "Flankline vs. Skyline." *Appalachia* 20, no. 4 (1934): 104–8.

———. "The Townless Highway." *The New Republic*, March 12, 1930, 93–95.

MacKaye, Benton, and Lewis Mumford. "Townless Highways for the Motorist: A Proposal for the Automobile Age." *Harper's Magazine*, August 1931, 347–56.

Mair, Kurt. *Die Hochstraßen der Alpen. Ein Autoführer über die Paß- und Gipfelstraßen in allen Teilen der Alpen*, 9th ed. Braunschweig: Richard Carl Schmidt, 1965.

Marx, Karl, and Friedrich Engels, *Manifesto of the Communist Party, The Marx-Engels Reader*, edited by Robert C. Tucker. New York: Norton, 1972.

Merrick, Hugh. *The Great Motor Highways of the Alps*. London: Robert Hale, 1958.

Michahelles, A.[ugust]. "Die Deutsche Alpenstraße." *VDI. Zeitschrift des Vereins Deutscher Ingenieure* 82, no. 37 (September 10, 1938): 1067–1071.

Mittermeier, Werner, and Albert Hirschbichler. *Traumlandschaften zwischen Bodensee und Berchtesgaden entlang der Deutschen Alpenstraße*. Berchtesgaden: Anton Plenk, 2003.

Müller-Alfeld, Theodor. *Die Deutsche Alpenstrasse: Vom Bodensee zum Königssee. In 122 Fotos.* Berlin: Stapp, 1970.

Mumford, Lewis. *The Myth of the Machine: The Pentagon of Power.* New York: Harcourt Brace Jovanovich, 1970.

———. *Technics and Civilization.* San Diego: Harcourt Brace Jovanovich, 1963 [1934].

Nash, Roderick. *The American Environment: Readings in the History of Conservation.* Reading, MA: Addison-Wesley, 1968.

National Park Service. *Blue Ridge Parkway, Virginia-North Carolina.* Washington, DC: Government Printing Office, 1947.

Nolen, John, and Henry V. Hubbard. *Parkways and Land Values.* Harvard City Planning Studies. vol. 11. Cambridge, MA: Harvard University Press, 1937.

Oberste Baubehörde Bayern. *Die Bayerischen Staatsstraßen. Die Ursachen ihres jetzigen schlechten Zustandes und die notwendigen Maßnahmen zu ihrer Verbesserung.* Munich: Wolf, 1925.

Ochi, Diane. Columbia River Highway Project. *Columbia River Highway: Options for Conservation and Reuse,* 1981. http://npshistory.com/publications/columbia-river-highway.pdf, accessed October 1, 2019.

Permanent International Association of Road Congresses. *Sixth International Road Congress, Washington, DC, 1930: Proceedings of the Congress.* Washington, DC: Government Printing Office, 1931.

Pirsig, Robert M. *Zen and the Art of Motorcycle Maintenance: An Inquiry into Values.* New York: Morrow, 1974.

Post, Emily. *By Motor to the Golden Gate.* New York: D. Appleton, 1916. https://archive.org /details/bymotortogoldengoopostiala, accessed October 1, 2019.

Prager, Christian. *Deutsche Alpenstrasse.* Hamm/Leipzig: Artcolor, 1996.

Public Roads Administration. *Toll Roads and Free Roads. Message from the President of the United States Transmitting a Letter from the Secretary of Agriculture, Concurred in by the Secretary of War, Enclosing a Report of the Bureau of Public Roads, United States Department of Agriculture, on the Feasibility of a System of Transcontinental Toll Roads and a Master Plan for Free Highway Development.* Washington, DC: Government Printing Office, 1939.

Reismann, Otto, ed. *Reichsautobahnen: Vom ersten Spatenstich zur fertigen Fahrbahn.* Berlin: Naturkunde und Technik Verlag F. Knapp, 1935.

Rives, Margaret Rose. *Blue Ridge Parkway.* Las Vegas, NV: KC Publications, 1982.

The Roads and Railroads, Vehicles, and Modes of Travelling, of Ancient and Modern Countries; with Accounts of Bridges, Tunnels, and Canals, in Various Parts of the World. London: J. W. Parker, 1839.

The Rossfeld as a film set. http://www.rossfeldpanoramastraße.de/informationen/das -rossfeld-alsfilmkulisse/, accessed April 23, 2013.

"Schloss Neuschwanstein heute—Besucherrekorde und Erhaltungsprobleme." https://www .neuschwanstein.de/deutsch/schloss/index.htm, accessed January 5, 2022.

Schmithals, Hans. *Die Deutsche Alpenstraße.* Berlin: Volk und Reich, 1936.

———. *Die Deutsche Alpenstraße. Vom Bodensee zum Königssee.* Lindau: Jan Thorbecke, 1960.

Schöner, Helmut. *Rund um den Watzmann. Streifzüge durch die Berchtesgadener Alpen.* Salzburg: Bergland-Buch, 1959.

Schwenkel, Hans. "Motorisierung und Naturschutz." *Kosmos* 35, no. 6 (1938): 201.

Seifert, Alwin. *Ein Leben für die Landschaft.* Düsseldorf: Diederichs, 1962.

Shelton-Roberts, Cheryl, and Bruce Roberts. *The Blue Ridge Parkway.* Morehead City, NC: Lighthouse Publications, 2002.

Simonson, Wilbur H. *Landscape Design and Its Relation to the Modern Highway, Lecture at Rutgers University, College of Engineering,* 1952. http://www.fhwa.dot.gov/infrastructure /simonson.cfm, accessed March 26, 2010.

Simonson, Wilbur H., and R. E. Royall. *Landschaftsgestaltung an der Straße. Roadside Improvement. Gekürzte Wiedergabe einer Veröffentlichung des U.S. Department of Agriculture Bureau of Public Roads.* Berlin: Volk und Reich, 1935.

Sims, Ralph, et al. "Transport." In *Climate Change 2014: Mitigation of Climate Change. Contribution of Working Group III to the Fifth Assessment Report of the Intergovernmental Panel on Climate Change,* edited by Ottmar Edenhofer et al., 599–670. Cambridge: Cambridge University Press, 2014. https://www.ipcc.ch/site/assets/uploads/2018/02/ipcc _wg3_ar5_chapter8.pdf.

Smithson, Alison. *AS in DS: An Eye on the Road.* Baden, Switzerland: Lars Müller, 2001.

Snow, W. Brewster, ed. *The Highway and the Landscape.* New Brunswick, NJ: Rutgers University Press, 1959.

Speck, Artur. *Via Vita: Lebensgeschichte eines Straßenbauers im Zeitalter des Kraftwagens.* Bad Godesberg: Kirschbaum Verlag, 1964.

Stephen, Leslie. *The Playground of Europe.* New York: Putnam, 1909. http://archive.org/stream /playgroundofeuroostepiala#page/372/mode/2up, accessed July 3, 2013.

Strache, Wolf. *Autoparadies Queralpenstrasse.* Stuttgart: Die schönen Bücher, 1955.

———. *Die Romantische Strasse. Mit einer Einführung und einer Karte.* Stuttgart: Die schönen Bücher, 1965.

Todt, Fritz. "Der nordische Mensch und der Verkehr." *Die Straße* 4 (1937): H. 14, S. 394–400.

Tunnard, Christopher, and Boris Pushkarev. *Man-Made America: Chaos or Control?* New Haven: Yale University Press, 1963.

United States Department of Commerce. *A Proposed Program for Scenic Roads & Parkways.* Washington, DC: Government Printing Office, 1966.

United States Environmental Protection Agency. "Carbon Pollution from Transportation." https://www.epa.gov/transportation-air-pollution-and-climate-change/carbon-pollution -transportation, accessed June 1, 2021.

United States Federal Highway Administration. *An Assessment of the Feasibility of Developing a National Scenic Highway System:* Report to Congress. Washington, DC: United States Department of Transportation, 1974.

United States Department of Transportation. Federal Highway Administration. *Scenic Byways.* Washington, DC: Federal Highway Administration, 1988.

Urban Advisors to the Federal Highway Administrator. "The Freeway in the City: Principles of Planning and Design: A Report to the Secretary, Dept. Of Transportation," edited by Department of Transportation. Washington, DC: Government Printing Office, 1968.

Venturi, Robert, Denise Scott Brown, and Steven Izenour. *Learning from Las Vegas*. Cambridge, MA: MIT Press, 1972.

Vilbig, [Josef] Ministerialrat. "VI. Internationaler Straßenkongreß in Washington." *ADAC Motorwelt* 28, no. 1 (1931): 20–22.

Walker, M. Lorenzo. "DeHaven Hinkson, M.D., 1891." *Journal of the National Medical Association* 66 (1974): 339–42.

Wells, Heber M. *See Europe If You Will, but See America First. An Address Delivered by Hon. Heber M. Wells before the "See America First" Conference at Salt Lake City, Utah, January 25th, 1906*. Denver: Carson-Harper, 1906. http://hdl.loc.gov/loc.gdc/scd0001.00160921481, accessed February 12, 2010.

Wojtowicz, Robert, ed. *Sidewalk Critic: Lewis Mumford's Writings on New York*. New York: Princeton Architectural Press, 1998.

Secondary Sources

Achenbach, Joel. *The Grand Idea: George Washington's Potomac and the Race to the West*. New York: Simon and Schuster, 2004.

Ahonen, Pertti. *After the Expulsion. West Germany and Eastern Europe, 1945–1990*. Oxford: Oxford University Press, 2003.

Akbal, Yasmin. "'Trinkt, O Augen, was die Wimper hält': Zur Biographie der Schwarzwaldhochstraße der 1920er bis 1970er Jahre." M.A. thesis, University of Mannheim, 2011.

Amato, Joseph Anthony. *On Foot: A History of Walking*. New York: New York University Press, 2004.

Amstädter, Rainer. *Der Alpinismus: Kultur, Organisation, Politik*. Vienna: WUV-Universitätsverlag, 1996.

Anderson, Larry. *Benton MacKaye: Conservationist, Planner, and Creator of the Appalachian Trail*. Creating the North American Landscape. Baltimore: Johns Hopkins University Press, 2002.

Andrews, Thomas G. "'Made by Toile'? Tourism, Labor, and the Construction of the Colorado Landscape, 1858–1917." *Journal of American History* 92, no. 3 (2005): 837–63.

Anker, Peder. *From Bauhaus to Ecohouse: A History of Ecological Design*. Baton Rouge: Louisiana State University Press, 2010.

Annese, Domenico. "Gilmore Clarke." In *Pioneers of American Landscape Design*, edited by Charles Birnbaum and Robin Karson, 56–60. New York: McGraw Hill, 2000.

Applegate, Celia. *A Nation of Provincials: The German Idea of Heimat*. Berkeley: University of California Press, 1990.

Asendorf, Christoph. *Super Constellation. Flugzeug und Raumrevolution. Die Wirkung der Luftfahrt auf Kunst und Kultur der Moderne*. Vienna: Springer, 1997.

Autry, Robyn, and Daniel J. Walkowitz. "Undoing the Flaneur." *Radical History Review* no. 114 (2012): 1–5.

Avila, Eric. *The Folklore of the Freeway: Race and Revolt in the Modernist City*. A Quadrant Book. Minneapolis: University of Minnesota Press, 2014.

Badenoch, Alexander. "Touring between War and Peace: Imagining the 'Transcontinental Motorway,' 1930–1950." *Journal of Transport History* 28, no. 2 (2007): 192–210.

Ballon, Hilary, and Kenneth T. Jackson, eds. *Robert Moses and the Modern City: The Transformation of New York*. New York: Norton, 2007.

Baranowski, Shelley, and Ellen Furlough, eds. *Being Elsewhere: Tourism, Consumer Culture, and Identity in Modern Europe and North America*. Ann Arbor: University of Michigan Press, 2001.

Barker, Theo. "A German Centenary in 1986, a French in 1995 or the Real Beginnings about 1905?" In *The Economic and Social Effects of the Spread of Motor Vehicles*, edited by Theo Barker, 1–54. London: Macmillan, 1987.

Barker, Theo, and Dorian Gerhold. *The Rise and Rise of Road Transport, 1700–1990*. Studies in Economic and Social History. Houndmills, Basingstoke, Hampshire: Macmillan, 1993.

Barnett, Gabrielle R. "Drive-by Viewing: Visual Consciousness and Forest Preservation in the Automobile Age." *Technology and Culture* 45, no. 1 (2004): 30–54.

Bätzing, Werner. *Die Alpen: Geschichte und Zukunft einer europäischen Kulturlandschaft*. 4th ed. Munich: Beck, 2015.

Baumann, Margret. "Auf den Spuren der deutschen Alpenpost Berchtesgaden-Lindau 2004: 520 Kilometer Autofahrt quer durch die Alpen." *Das Archiv. Magazin für Post- und Telekommunikationsgeschichte* no. 3 (2004): 60–66.

Bayerischer Rundfunk. *Vier Jahrhunderte Ringen mit der Passion*. http://www.br.de/themen /bayern/passionsspiele-oberammergau100.html, accessed April 24, 2013.

Beierl, Florian M. *Geschichte des Kehlsteins. Ein Berg verändert sein Gesicht*. 7th ed. Berchtesgaden: Anton Plenk, 2004.

Belasco, Warren James. *Americans on the Road: From Autocamp to Motel, 1910–1945*. Baltimore: Johns Hopkins University Press, 1979.

Benesch, Klaus. "The Dynamic Sublime: Geschwindigkeit und Ästhetik in der amerikanischen Moderne." In *Raum- und Zeitreisen: Studien zur Literatur und Kultur des 19. und 20. Jahrhunderts*, edited by Hans Ulrich Seeber and Julika Griem, 102–17. Tübingen: Niemeyer, 2003.

Berghoff, Hartmut, and Uwe Spiekermann, eds. *Decoding Modern Consumer Societies*. New York: Palgrave Macmillan, 2012.

Bermingham, Ann. *Landscape and Ideology: The English Rustic Tradition, 1740–1850*. Berkeley: University of California Press, 1986.

Bigg, Charlotte. "The Panorama, or La Nature à Coup d'Oeil." In *Observing Nature—Representing Experience. The Osmotic Dynamics of Romanticism 1800–1850*, edited by Erna Fiorentini, 73–95. Berlin: Reimer, 2007.

Bijsterveld, Karin, Eefje Cleophas, Stefan Krebs, and Gijs Mom. *Sound and Safe: A History of Listening behind the Wheel*. New York: Oxford University Press, 2014.

Black, Brian. "Organic Planning: Ecology and Design in the Landscape of the Tennessee Valley Authority, 1933–45." In *Environmentalism in Landscape Architecture*, edited by Michel Conan. Dumbarton Oaks Colloquium on the History of Landscape Architecture, 71–96. Washington, DC: Dumbarton Oaks, 2000.

Blackbourn, David. *The Conquest of Nature: Water, Landscape, and the Making of Modern Germany*. New York: Norton, 2006.

Blackbourn, David, and James N. Retallack, eds. *Localism, Landscape, and the Ambiguities of*

Place: German-Speaking Central Europe, 1860–1930. Toronto: University of Toronto Press, 2007.

Blanke, David. *Hell on Wheels. The Promise and Peril of America's Car Culture, 1900–1940*. Lawrence: University of Kansas Press, 2007.

Blaszczyk, Regina Lee. *American Consumer Society, 1865–2005: From Hearth to HDTV*. Hoboken, NJ: Wiley, 2009.

Borscheid, Peter. *Das Tempo-Virus: Eine Kulturgeschichte der Beschleunigung*. Frankfurt am Main: Campus, 2004.

Bright, Randy. *Disneyland: Inside Story*. New York: Abrams, 1987.

Brown, Kate. *Plutopia: Nuclear Families, Atomic Cities, and the Great Soviet and American Nuclear Disasters*. New York: Oxford University Press, 2013.

Brüggemeier, Franz-Josef, Mark Cioc, and Thomas Zeller, eds. *How Green Were the Nazis? Nature, Environment, and Nation in the Third Reich*. Series in Ecology and History. Athens: Ohio University Press, 2005.

Burckhardt, Lucius. *Why Is Landscape Beautiful? The Science of Strollology*. Basel: Birkhäuser, 2015.

Campanella, Thomas J. "American Curves: Gilmore D. Clarke and the Modern Civic Landscape." *Harvard Design Magazine*, Summer 1997, 40–43.

Caro, Robert A. *The Power Broker: Robert Moses and the Fall of New York*. New York: Knopf, 1974.

Carr, Ethan. *Mission 66: Modernism and the National Park Dilemma*. Amherst: University of Massachusetts Press, 2007.

———. *Wilderness by Design: Landscape Architecture and the National Park Service*. Lincoln: University of Nebraska Press, 1998.

Carse, Ashley. *Beyond the Big Ditch: Politics, Ecology, and Infrastructure at the Panama Canal*. Cambridge, MA: MIT Press, 2014.

Catton, Theodore. "The Hegemony of the Car Culture in U.S. National Parks." In *Public Nature: Scenery, History, and Park Design*, edited by Ethan Carr, Shaun Eyring, and Richard Guy Wilson. Charlottesville: University of Virginia Press, 2013.

Chamberlin, Silas. *On the Trail: A History of American Hiking*. New Haven: Yale University Press, 2016.

Chaney, Sandra. *Nature of the Miracle Years: Conservation in West Germany, 1956–1975*. New York: Berghahn, 2008.

Chaussy, Ulrich. *Nachbar Hitler. Führerkult und Heimatzerstörung am Obersalzberg*. 7th ed. Berlin: Links, 2012.

Clarke, Deborah. *Driving Women: Fiction and Automobile Culture in Twentieth-Century America*. Baltimore: Johns Hopkins University Press, 2007.

Clarsen, Georgine. *Eat My Dust: Early Women Motorists*. Baltimore: Johns Hopkins University Press, 2008.

Coleman, Simon, and Mike Crang, eds. *Tourism: Between Place and Performance*. New York: Berghahn, 2002.

Confino, Alon. *The Nation as a Local Metaphor: Württemberg, Imperial Germany, and National Memory, 1871–1918*. Chapel Hill: University of North Carolina Press, 1997.

———. "Traveling as a Culture of Rememberance: Traces of National Socialism in West Germany, 1945–1960." *History and Memory* 12, no. 2 (2000): 92–121.

Corn, Joseph J. *User Unfriendly: Consumer Struggles with Personal Technologies, from Clocks and Sewing Machines to Cars and Computers.* Baltimore: Johns Hopkins University Press, 2011.

———. *Winged Gospel: America's Romance with Aviation.* Baltimore: Johns Hopkins University Press, 2002.

Corn, Joseph J., Brian Horrigan, and Katherine Chambers. *Yesterday's Tomorrows: Past Visions of the American Future.* New York: Summit Books, 1984.

Crary, Jonathan. *Techniques of the Observer: On Vision and Modernity in the Nineteenth Century.* Cambridge, MA: MIT Press, 1990.

Cresswell, Tim. "Towards a Politics of Mobility." *Environment and Planning D: Society and Space* 28, no. 1 (2010): 17–31.

Cronon, William: "The Trouble with Wilderness; or, Getting Back to the Wrong Nature." In *Uncommon Ground: Rethinking the Human Place in Nature,* edited by William Cronon, 69–90. New York: Norton, 1995.

———, ed. *Uncommon Ground: Rethinking the Human Place in Nature.* New York: Norton, 1995.

Cushman, Gregory T. "Environmental Therapy for Soil and Social Erosion: Landscape Architecture and Depression-Era Highway Construction in Texas." In *Environmentalism in Landscape Architecture,* edited by Michel Conan, 45–70. Dumbarton Oaks Colloquium on the History of Landscape Architecture. Washington, DC: Dumbarton Oaks, 2000.

Cutler, Phoebe. *The Public Landscape of the New Deal.* New Haven: Yale University Press, 1985.

Dalbey, Matthew. *Regional Visionaries and Metropolitan Boosters: Decentralization, Regional Planning, and Parkways during the Interwar Years.* Boston: Kluwer, 2002.

Daniels, Stephen. *Humphry Repton: Landscape Gardening and the Geography of Georgian England.* New Haven: Yale University Press, 1999.

———. "Marxism, Culture, and the Duplicity of Landscape." In *New Models in Geography: The Political-Economy Perspective,* edited by Richard Peet and Nigel Thrift, 162–79. London: Unwin Hyman, 1989.

Davis, Donald Edward. *Where There Are Mountains.* Athens: University of Georgia Press, 2000.

Davis, George T. "Protecting Scenic Views: Seventy Years of Managing and Enforcing Scenic Easements along the Blue Ridge Parkway." MA thesis. Virginia Polytechnic Institute and State University, 2009.

Davis, Timothy. "The Bronx River Parkway and Photography as an Instrument of Landscape Reform." *Studies in the History of Gardens & Designed Landscapes* 27 (2007): 113–41.

———. "Looking Down the Road: J. B. Jackson and the American Highway." In *Everyday America Cultural Landscape Studies after J. B. Jackson,* edited by Chris Wilson and Paul Groth, 62–80. Berkeley: University of California Press, 2003.

———. *National Park Roads: A Legacy in the American Landscape.* Charlottesville: University of Virginia Press, 2016.

Davis, Timothy F. "The American Parkway as Colonial Revival Landscape." In *Re-Creating the American Past: Essays on the Colonial Revival*, edited by Richard Guy Wilson, Shaun Eyring, and Kenny Marotta, 140–66. Charlottesville: University of Virginia Press, 2006.

———. "Mount Vernon Memorial Highway and the Evolution of the American Parkway." PhD dissertation. University of Texas, 1997.

———. "Mount Vernon Memorial Highway: Changing Conceptions of an American Commemorative Landscape." In *Places of Commemoration: Search for Identity and Landscape Design*, edited by Joachim Wolschke Bulmahn, 123–77. Washington, DC: Dumbarton Oaks, 2001.

———. "The Rise and Decline of the American Parkway." In *The World beyond the Windshield: Roads and Landscapes in the United States and Europe*, edited by Christof Mauch and Thomas Zeller, 35–58. Athens: Ohio University Press, 2008.

De Grazia, Victoria. *Irresistible Empire: America's Advance through Twentieth-Century Europe.* Cambridge, MA: Belknap Press of Harvard University Press, 2005.

Denning, Andrew. *Skiing into Modernity: A Cultural and Environmental History.* Berkeley: University of California Press, 2014.

Desportes, Marc. *Paysages en mouvement: Transports et perception de l'espace, XVIIIe–XXe siècle.* Paris: Gallimard, 2005.

Dienel, Hans-Liudger, and Hans-Ulrich Schiedt, eds. *Die moderne Straße: Planung, Bau und Verkehr vom 18. bis zum 20. Jahrhundert.* Frankfurt am Main: Campus, 2010.

Dilsaver, Lary M. *America's National Park System: The Critical Documents.* 2nd ed. Lanham, MD: Rowman & Littlefield, 2016.

Dilsaver, Lary M., and William C. Tweed. *Challenge of the Big Trees: A Resource History of Sequoia and Kings Canyon National Parks.* Three Rivers, CA: Sequoia Natural History Association, 1990. https://www.nps.gov/parkhistory/online_books/dilsaver-tweed/chap6d.htm, accessed May 26, 2021.

Divall, Colin. "Mobilities and Transport History." In *The Routledge Handbook of Mobilities*, edited by Peter Adey, David Bissell, Kevin Hannam, Peter Merriman, and Mimi Sheller, 36–44. London: Routledge, 2014.

Doss, Erika. "Making Imagination Safe in the 1950s: Disneyland's Fantasy Art and Architecture." In *Designing Disney's Theme Parks: The Architecture of Reassurance*, edited by Karal Ann Marling, 179–89. New York: Flammarion, 1997.

Dümpelmann, Sonja. *Flights of Imagination: Aviation, Landscape, Design.* Charlottesville: University of Virginia Press, 2014.

Dunlavy, Colleen A. *Politics and Industrialization: Early Railroads in the United States and Prussia.* Princeton: Princeton University Press, 1994.

Écomusée du pays de la Roudoule. *La route des grandes alpes.* Puget-Rostang: Édition de l'écomusée du pays de la Roudoule, 2008.

Edwards, Paul N. "Infrastructure and Modernity: Force, Time, and Social Organization in the History of Sociotechnical Systems." In *Modernity and Technology*, edited by Thomas J. Misa, Philip Brey, and Andrew Feenberg, 185–225. Cambridge, MA: MIT Press, 2003.

Eisenfeld, Sue. *Shenandoah: A Story of Conservation and Betrayal.* Lincoln: University of Nebraska Press, 2014.

Ekström, Anders. "Seeing from Above: A Particular History of the General Observer." *Nineteenth-Century Contexts* 31, no. 3 (2009): 185–207.

Engels, Jens-Ivo. *Naturpolitik in der Bundesrepublik. Ideenwelt und politische Verhaltensstile in Naturschutz und Umweltbewegung 1950–1980*. Paderborn: Schöningh, 2006.

Esperdy, Gabrielle. *American Autopia: An Intellectual History of the American Roadside at Midcentury*. Charlottesville: University of Virginia Press, 2019.

Fabian, Sina. *Boom in der Krise: Konsum, Tourismus, Autofahren in Westdeutschland und Großbritannien 1970–1990*. Göttingen: Wallstein, 2016.

Fahl, Ronald J. "S. C. Lancaster and the Columbia River Highway: Engineer as Conservationist." *Oregon Historical Quarterly* 74, no. 2 (1973): 101–44.

Fein, Michael R. *Paving the Way: New York Roadbuilding and the American State, 1880–1956*. Lawrence: University of Kansas Press, 2008.

Finney, Carolyn. *Black Faces, White Spaces: Reimagining the Relationship of African Americans to the Great Outdoors*. Chapel Hill: University of North Carolina Press, 2014.

Flik, Reiner. *Von Ford Lernen? Automobilbau und Motorisierung in Deutschland bis 1933*. Wirtschafts- und Sozialhistorische Studien 11. Cologne: Böhlau, 2001.

Flonneau, Mathieu. *Les cultures du volant. Essaie sur les mondes de l'automobilisme, XXe–XXIe siècles*. Paris: Éditions Autrement, 2008.

Frank, Alison F. "The Air Cure Town: Commodifying Mountain Air in Alpine Central Europe." *Central European History* 45 (2012): 185–207.

Fraunholz, Uwe. *Motorphobia: Anti-Automobiler Protest in Kaiserreich und Weimarer Republik*. Göttingen: Vandenhoeck & Ruprecht, 2002.

Freytag, Anette. "When the Railway Conquered the Garden: Velocity in Parisian and Viennese Parks," In *Landscape Design and the Experience of Motion*, edited by Michel Conan, 215–42. Dumbarton Oaks Colloquium on the History of Landscape Architecture XXIV, Washington, DC, 2003.

Fritzsche, Peter. *A Nation of Fliers: German Aviation and the Popular Imagination*. Cambridge, MA: Harvard University Press, 1992.

Furter, Reto. "Hintergrund des Alpendiskurses: Indikatoren und Karten." In *Die Alpen! Les Alpes! Zur europäischen Wahrnehmungsgeschichte seit der Renaissance*, edited by Jon Mathieu and Simona Boscani Leoni, 73–96. Bern: Peter Lang, 2005.

Gabriel, Roland. *Dem Auto eine Bahn. Deutsche "Nurautostraßen" vor 1933*. Archiv für die Geschichte des Straßen- und Verkehrswesens. vol. 23, Cologne: Forschungsgesellschaft für Straßen- und Verkehrswesen, 2010.

Gall, Alexander. *"Gute Straßen bis ins kleinste Dorf!" Verkehrspolitik in Bayern zwischen Wiederaufbau und Ölkrise*. Frankfurt am Main: Campus, 2005.

———. "Straßen und Straßenverkehr (19./20. Jahrhundert)." *Historisches Lexikon Bayerns*. https://www.historisches-lexikon-bayerns.de/Lexikon/ Straßen und Straßenverkehr _(19./20._Jahrhundert), accessed August 15, 2021.

Gandy, Matthew. *Concrete and Clay: Reworking Nature in New York City*. Cambridge, MA: MIT Press, 2002.

Garraty, John A. "The New Deal, National Socialism, and the Great Depression." *American Historical Review* 78, no. 4 (1973): 907–44.

Gassert, Philipp. *Amerika im Dritten Reich: Ideologie, Propaganda und Volksmeinung 1933–1945*. Stuttgart: Franz Steiner, 1997.

Gelberg, Karl-Ulrich. "Die Oberste Baubehörde zwischen 1932 und 1949. Zur Kontinuität einer bayerischen Zentralbehörde." In *Staat und Gau in der NS-Zeit. Bayern 1933–1945*, edited by Hermann Rumschöttel and Walter Ziegler, 297–339. Munich: Beck, 2004.

Gillespie, Angus Kress, and Michael Aaron Rockland. *Looking for America on the New Jersey Turnpike*. New Brunswick, NJ: Rutgers University Press, 1992.

Gioielli, Robert. *Environmental Activism and the Urban Crisis: Baltimore, St. Louis, Chicago*. Philadelphia: Temple University Press, 2014.

Greene, Ann Norton. *Horses at Work: Harnessing Power in Industrial America*. Cambridge, MA: Harvard University Press, 2008.

Gregg, Sara M. *Managing the Mountains: Land Use Planning, the New Deal, and the Creation of a Federal Landscape in Appalachia*. New Haven: Yale University Press, 2010.

Gudis, Catherine. *Buyways: Billboards, Automobiles and the American Landscape*. New York: Routledge, 2004.

Guldi, Jo. *Roads to Power: Britain Invents the Infrastructure State*. Cambridge, MA: Harvard University Press, 2012.

Günther, Dagmar. *Alpine Quergänge: Kulturgeschichte des bürgerlichen Alpinismus, 1870–1930*. Frankfurt am Main: Campus, 1998.

Guse, John C. "Nazi Technical Thought Revisited." *History and Technology* 26, no. 1 (2010): 3–33.

Gutfreund, Owen D. *Twentieth Century Sprawl: Highways and the Reshaping of the American Landscape*. New York: Oxford University Press, 2004.

Hachtmann, Rüdiger. *Tourismus-Geschichte*. Göttingen: UTB, 2007.

Hall, Marcus. *Earth Repair: A Transatlantic History of Environmental Restoration*. Charlottesville: University of Virginia Press, 2005.

———, ed. *Restoration and History: The Search for a Usable Environmental Past*. New York: Routledge, 2010.

Hamilton, Shane. *Trucking Country: The Road to America's Wal-Mart Economy*. Politics and Society in Twentieth-Century America. Princeton: Princeton University Press, 2008.

Hansen, Peter H. *The Summits of Modern Man: Mountaineering after the Enlightenment*. Cambridge, MA: Harvard University Press, 2013.

Haraway, Donna. "Situated Knowledges. The Science Question in Feminism and the Privilege of Partial Perspective." *Feminist Studies* 14, no. 3 (1988): 575–99.

Hasenöhrl, Ute. *Zivilgesellschaft und Protest: Eine Geschichte der Naturschutz- und Umweltbewegung in Bayern 1945–1980*. Göttingen: Vandenhoeck und Ruprecht, 2011.

Haslauer, Johannes. "Kesselbergstraße." *Historisches Lexikon Bayerns*. http://www.historisches-lexikon-bayerns.de/Lexikon/Kesselbergstraße, accessed January 5, 2022.

Havlick, David. *No Place Distant. Roads and Motorized Recreation on America's Public Lands*. Washington, DC: Island Press, 2002.

Heckmann-Strohkark, Ingrid. "Der Traum von einer europäischen Gemeinschaft: Die internationalen Autobahnkongresse 1931 und 1932." In *Die Schweizer Autobahn*, edited by Martin Heller and Andreas Volk, 32–45. Zürich: Museum für Gestaltung, 1999.

Heißerer, Dirk. *Ludwig II*. Reinbek: Rowohlt, 2003.

Heitmann, John Alfred. *The Automobile and American Life*. Jefferson, NC: McFarland, 2009.

Hochstetter, Dorothee. *Motorisierung und "Volksgemeinschaft": Das Nationalsozialistische Kraftfahrkorps, NSKK, 1931–1945*. Studien zur Zeitgeschichte. München: R. Oldenbourg, 2005.

Hokanson, Drake, and Douglas Pappas. *The Lincoln Highway: Main Street across America*. Iowa City: University of Iowa Press, 1988.

Hölzl, Richard. "Naturschutz in Bayern von 1905–1945: Der Landesausschuß für Naturpflege und der Bund für Naturschutz zwischen privater und staatlicher Initiative." MA thesis, University of Regensburg, 2005. http://www.opus-bayern.de/uni-regensburg /volltexte/2005/521/pdf/RDTGKU1.pdf, accessed April 5, 2010.

Houston, Kerr. "'A Proper Medium': Early Motorists' Perception of the European Landscape." *Early Popular Visual Culture* 7, no. 1 (2009): 29–43.

Hughes, Thomas P. *American Genesis: A Century of Innovation and Technological Enthusiasm, 1870–1970*. New York: Viking, 1989.

Hughes, Thomas P., and Agatha C. Hughes, eds. *Lewis Mumford: Public Intellectual*. New York: Oxford University Press, 1990.

Hunt, John Dixon. *Gardens and the Picturesque: Studies in the History of Landscape Architecture*. Cambridge, MA: MIT Press, 1992.

Hvattum, Mari, Janike Kampevold Larsen, Brita Brenna, and Beate Elvebakk, eds. *Routes, Roads and Landscapes*. Farnham: Ashgate, 2011.

Interrante, Joseph. "You Can't Go to Town in a Bathtub: Automobile Movement and the Reorganization of Rural American Space, 1900–1930." *Radical History Review* 21 (1979): 151–68.

Jacoby, Karl. *Crimes against Nature: Squatters, Poachers, Thieves, and the Hidden History of American Conservation*. Berkeley: University of California Press, 2001.

Jakle, John A. *The Tourist: Travel in Twentieth-Century North America*. Lincoln: University of Nebraska Press, 1985.

Jakle, John A., and Keith A. Sculle. *Motoring: The Highway Experience in America*. Athens: University of Georgia Press, 2008.

Joerges, Bernward. "Do Politics Have Artefacts?" *Social Studies of Science* 29 (1999): 411–31.

Kahrl, Andrew W. *The Land Was Ours: African American Beaches from Jim Crow to the Sunbelt South*. Chapel Hill: University of North Carolina Press, 2016.

Kania-Schütz, Monika, ed. *Die Deutsche Alpenstraße. Deutschlands älteste Ferienroute*. Munich: Volk, 2021.

———. "Einleitung." In *Die Deutsche Alpenstraße. Deutschlands älteste Ferienroute*, edited by Monika Kania-Schütz, 9–22. Munich: Volk, 2021.

Kaschuba, Wolfgang. *Die Überwindung der Distanz: Zeit und Raum in der europäischen Moderne*. Frankfurt am Main: Fischer, 2004.

Kates, James. "A Path Made of Words: The Journalistic Construction of the Appalachian Trail." *American Journalism* 30, no. 1 (2013): 112–34.

Keller, Tait. *Apostles of the Alps: Mountaineering and Nation Building in Germany and Austria, 1860–1939*. Chapel Hill: University of North Carolina Press, 2016.

———. "The Mountains Roar: The Alps during the Great War." *Environmental History* 13, no. 2 (2009): 253–74.

Kemp, Louis Ward. "Aesthetes and Engineers. The Occupational Ideology of Highway Design." *Technology and Culture* 27 (1986): 759–97.

Kern, Stephen. *The Culture of Time and Space, 1880–1918.* Cambridge, MA: Harvard University Press, 1983.

Kershaw, Ian. *The 'Hitler Myth': Image and Reality in the Third Reich.* Oxford: Oxford University Press, 2001.

Kirsch, David A. *The Electric Vehicle and the Burden of History.* New Brunswick, NJ: Rutgers University Press, 2000.

Kleinschmidt, Christian. *Technik und Wirtschaft im 19. und 20. Jahrhundert.* Munich: Oldenbourg, 2006.

Klenke, Dietmar. "Autobahnbau in Westfalen von den Anfängen bis zum Höhepunkt der 1970er Jahre. Eine Geschichte der politischen Planung." In *Verkehr und Region im 19. und 20. Jahrhundert. Westfälische Beispiele,* edited by Wilfried Reininghaus and Karl Teppe, 249–70. Paderborn: Schöningh, 1999.

———. *Bundesdeutsche Verkehrspolitik und Motorisierung. Konfliktträchtige Weichenstellungen in den Jahren des Wiederaufstiegs.* Zeitschrift für Unternehmensgeschichte, Beiheft. vol. 79. Stuttgart: Steiner, 1993.

———. *Freier Stau für freie Bürger: Die Geschichte der bundesdeutschen Verkehrspolitik 1949–1994.* Darmstadt: Wissenschaftliche Buchgesellschaft, 1995.

Kline, Ronald R. *Consumers in the Country: Technology and Social Change in Rural America.* Revisiting Rural America. Baltimore: Johns Hopkins University Press, 2000.

Klösch, Christian. "The Great Auto Theft: Confiscation and Restitution of Motorised Vehicles in Austria during and after the Nazi Period." *Journal of Transport History* 34, no. 2 (2013): 140–61.

Kocka, Jürgen. *Arbeiterleben und Arbeiterkultur: Die Entstehung einer sozialen Klasse.* Bonn: Dietz, 2015.

König, Gudrun M. *Eine Kulturgeschichte des Spaziergangs. Spuren einer bürgerlichen Praktik 1780–1850.* Vienna: Böhlau, 1996.

König, Sebald. "Die Roßfeld-Panoramastraße, einst und heute." *Zeitschrift der Vereinigung der Straßenbau- und Verkehrsingenieure in Bayern,* 4–10. 2004. http://www.vsvi-bayern.de/?loadCustomFile=30_Downloads/Zeitschrift/2004_Zeitschrift.pdf, accessed April 23, 2013.

König, Wolfgang. "Adolf Hitler vs. Henry Ford: The *Volkswagen,* the Role of America as a Model, and the Failure of Nazi Consumer Society." *German Studies Review* 27, no. 2 (2004): 249–68.

———. *Bahnen und Berge: Verkehrstechnik, Tourismus und Naturschutz in den Schweizer Alpen 1870–1939.* Frankfurt am Main: Campus, 2000.

———. *Volkswagen, Volksempfänger, Volksgemeinschaft: "Volksprodukte" im Dritten Reich. Vom Scheitern einer nationalsozialistischen Konsumgesellschaft.* Paderborn: Schöningh, 2004.

Kopper, Christopher M. "The Breakthrough of the Package Tour in Germany after 1945." *Journal of Tourism History* 1, no. 1 (2009): 67–92.

Koshar, Rudy. *German Travel Cultures. Leisure, Consumption, and Culture.* Oxford: Berg, 2000.

———. "Germans at the Wheel: Cars and Leisure Travel in Interwar Germany." In *Histories of Leisure,* edited by Rudy Koshar, 215–30. Oxford: Berg, 2002.

————. "Organic Machines: Cars, Drivers, and Nature from Imperial to Nazi Germany." In *Germany's Nature: Cultural Landscapes and Environmental History*, edited by Thomas Lekan and Thomas Zeller, 111–39. New Brunswick, NJ: Rutgers University Press, 2005.

————. "'What Ought to Be Seen': Tourists' Guidebooks and National Identities in Modern Germany and Europe." *Journal of Contemporary History* 33, no. 3 (1998): 323–40.

Kotter, Simon. "Von Rennfahrern und Genussfahrern. Motorsport auf der Deutschen Alpenstraße." In *Die Deutsche Alpenstraße. Deutschlands älteste Ferienroute*, edited by Monika Kania-Schütz, 101–18. Munich: Volk, 2021.

Kreuzer, Bernd. "Die Deutsche Alpenstraße zwischen Natur, Tourismus, Automobilismus und Politik." In *Die Deutsche Alpenstraße. Deutschlands älteste Ferienroute*, edited by Monika Kania-Schütz, 63–82. Munich: Volk, 2021.

————. "'Dieser zahlungskräftige Ausländer aber macht seine Reisen vielfach nur mit dem Auto': Straßen als Voraussetzung und Attraktion für den modernen Alpentourismus." In *Alpenreisen. Erlebnis–Raumtransformationen–Imagination*, edited by Kurt Luger and Franz Rest, 167–84. Innsbruck: Studien Verlag, 2017.

Krige, John, ed. *How Knowledge Moves: Writing the Transnational History of Science and Technology*. Chicago: University of Chicago Press, 2019.

Krim, Arthur. *Route 66: Iconography of the American Highway*. Chicago: University of Chicago Press, 2006.

Kroll, Gary. "An Environmental History of Roadkill: Road Ecology and the Making of the Permeable Highway." *Environmental History* 20, no. 1 (2015): 4–28.

Laak, Dirk van. *Alles im Fluss. Die Lebensadern unserer Gesellschaft. Geschichte und Zukunft der Infrastruktur*. Frankfurt am Main: Fischer, 2018.

Ladd, Brian. *Autophobia: Love and Hate in the Automobile Age*. Chicago: University of Chicago Press, 2008.

Laderman, David. *Driving Visions: Exploring the Road Movie*. Austin: University of Texas Press, 2002.

LaFrank, Kathleen. "Real and Ideal Landscapes along the Taconic State Parkway." *Perspectives in Vernacular Architecture* 9 (2003): 247–62.

Larkin, Brian. "The Politics and Poetics of Infrastructure." *Annual Review of Anthropology* 42 (2013): 327–43.

Lavenir, Catherine Bertho. *La roue et le stylo. Comment nous sommes devenues touristes*. Paris: Odile Jacob, 1999.

Lay, Maxwell G. *Ways of the World: A History of the World's Roads and of the Vehicles That Used Them*. New Brunswick, NJ: Rutgers University Press, 1992.

Lekan, Thomas M. *Imagining the Nation in Nature: Landscape Preservation and German Identity, 1885–1945*. Cambridge, MA: Harvard University Press, 2003.

Lekan, Thomas, and Thomas Zeller. "Region, Scenery, and Power: Cultural Landscapes in Environmental History." In *The Oxford Handbook of Environmental History*, edited by Andrew C. Isenberg, 332–65. Oxford: Oxford University Press, 2014.

Lewis, Michael, ed. *American Wilderness: A New History*. New York: Oxford University Press, 2007.

Lewis, Tom. *Divided Highways: Building the Interstate Highways, Transforming American Life*. New York: Viking, 1997.

Lilla, Joachim. "Vilbig, Josef." Staatsminister, leitende Verwaltungsbeamte und (NS-) Funktionsträger in Bayern 1918 bis 1945. http://verwaltungshandbuch.bayerische -landesbibliothek-online.de/vilbig-josef, accessed March 18, 2013.

Lobenhofer-Hirschbold, Franziska. "Fremdenverkehr (Von den Anfängen bis 1945)." *Historisches Lexikon Bayerns*. http://www.historisches-lexikon-bayerns.de/artikel/artikel _44714, accessed April 24, 2013.

Louter, David. *Windshield Wilderness: Cars, Roads, and Nature in Washington's National Parks*. Seattle: University of Washington Press, 2006.

Lyth, Peter. "Supersonic/Gin and Tonic: The Rise and Fall of Concorde, 1950–2000." In *Transport and Its Place in History: Making the Connections*, edited by David Turner, 15–33. London: Routledge 2019.

Mackintosh, Barry. "Shootout on the Old C. & O. Canal: The Great Parkway Controversy, 1950–1960." *Maryland Historical Magazine* 90, no. 2 (Summer 1995): 140–63.

Maher, Neil. *Nature's New Deal: The Civilian Conservation Corps and the Roots of the American Environmental Movement*. New York: Oxford University Press, 2008.

Marchand, Roland. "The Designers Go to the Fair II: Norman Bel Geddes, the General Motors 'Futurama,' and the Visit to the Factory Transformed." *Design Issues* 8, no. 2 (1992): 22–40.

Mason, Randall. *The Once and Future New York: Historic Preservation and the Modern City*. Minneapolis: University of Minnesota Press, 2009.

Mathieu, Jon. "Alpenwahrnehmung: Probleme der historischen Periodisierung." In *Die Alpen! Les Alpes! Zur europäischen Wahrnehmungsgeschichte seit der Renaissance*, edited by Jon Mathieu and Simona Boscani Leoni, 53–72. Bern: Peter Lang, 2005.

———. *History of the Alps, 1500–1900: Environment, Development, and Society*. Morgantown: West Virginia University Press, 2009.

Mauch, Christof, and Thomas Zeller, eds. *The World beyond the Windshield: Roads and Landscapes in the United States and Europe*. Athens: Ohio University Press, 2008.

McCarthy, Tom. *Auto Mania: Cars, Consumers, and the Environment*. New Haven: Yale University Press, 2007.

McClelland, Linda Flint. *Building the National Parks: Historic Landscape Design and Construction*. Baltimore: Johns Hopkins University Press, 1998.

McConnell, Curt, and Douglas Pappas. *Coast to Coast by Automobile: The Pioneering Trips, 1899–1908*. Stanford, CA: Stanford University Press, 2000.

McCullough, Robert L. *Old Wheelways: Traces of Bicycle History on the Land*. Cambridge, MA: MIT Press, 2015.

McNeill, J. R. *Something New under the Sun: An Environmental History of the Twentieth-Century World*. New York: Norton, 2000.

McShane, Clay. *Down the Asphalt Path. The Automobile and the American City*. New York: Columbia University Press, 1994.

McShane, Clay, and Joel A. Tarr. *The Horse in the City: Living Machines in the Nineteenth Century*. Baltimore: Johns Hopkins University Press, 2007.

Merki, Christoph Maria. *Der holprige Siegeszug des Automobils 1895–1930: Zur Motorisierung des Strassenverkehrs in Frankreich, Deutschland und der Schweiz*. Vienna: Böhlau, 2002.

Merriman, Peter. *Driving Spaces: A Cultural-Historical Geography of England's M1 Motorway.* Malden, MA: Blackwell, 2007.

———. "Road Works: Some Observations on Representing Roads." *Transfers* 5, no. 1 (2015): 108–13.

———. "Roads: Lawrence Halprin, Modern Dance and the American Freeway Landscape." In *Geographies of Mobilities: Practices, Spaces, Subjects*, edited by Tim Cresswell and Peter Merriman, 99–118. Farnham: Ashgate, 2011.

Merriman, Peter, and Lynne Pearce. "Mobility and the Humanities." *Mobilities* 12, no. 4 (2017): 493–508.

Mierzejewski, Alfred C. *The Most Valuable Asset of the Reich: A History of the German Railway Company.* 2 vol. Chapel Hill: University of North Carolina Press, 1999/2000.

Miles, John C. *Wilderness in National Parks: Playground or Preserve.* Seattle: University of Washington Press, 2009.

Misa, Thomas J. *A Nation of Steel: The Making of Modern America, 1865–1925.* Baltimore: Johns Hopkins University Press, 1995.

Mitchell, Arthur H. *Hitler's Mountain: The Führer, Obersalzberg and the American Occupation of Berchtesgaden.* Jefferson, NC: McFarland, 2007.

Mittlefehldt, Sarah. *Tangled Roots: The Appalachian Trail and American Environmental Politics.* Seattle: University of Washington Press, 2013.

Mohl, Raymond A. "Stop the Road: Freeway Revolts in American Cities." *Journal of Urban History* 30, no. 5 (2004): 674–706.

Mölter, Max. *Die Hochrhönstraße. Geschichtliche, erdkundliche, erdgeschichtliche, naturkundlich, wirtschaftliche, kulturkundliche und volkskundliche Bemerkungen über eine Landschaft.* Fulda: Parzeller, 1986.

Mom, Gijs. *Atlantic Automobilism: Emergence and Persistence of the Car, 1895–1940.* New York: Berghahn, 2015.

———. "Civilized Adventure as a Remedy for Nervous Times: Early Automobilism and *Fin-de-Siècle* Culture." *History of Technology* 23 (2001): 157–90.

———. "Roads without Rails: European Highway-Network Building and the Desire for Long-Range Motorized Mobility." *Technology and Culture* 46, no. 4 (2005): 745–72.

Mom, Gijs P. A., and David A. Kirsch. "Technologies in Tension: Horses, Electric Trucks, and the Motorization of American Cities, 1900–1925." *Technology and Culture* 42, no. 3 (2001): 489–518.

Moraglio, Massimo. *Driving Modernity: Technology, Experts, Politics and Fascist Motorways, 1922–1943.* Translated by Erin O'Loughlin. New York: Berghahn, 2017.

———. "Transferring Technology, Shaping Society: Traffic Engineering in PIARC Agenda, in the Early 1930s." *Technikgeschichte* 80, no. 1 (2013): 13–32.

Möser, Kurt. "The Dark Side of 'Automobilism,' 1900–30: Violence, War and the Motor Car." *Journal of Transport History* 24, no. 2 (2003): 238–58.

———. *Geschichte des Autos.* Frankfurt am Main: Campus, 2002.

Mukerji, Chandra. *Impossible Engineering: Technology and Territoriality on the Canal du Midi.* Princeton: Princeton University Press, 2009.

Müller, Uwe. "Der Beitrag des Chausseebaus zum Modernisierungsprozess in Preußen." In

Die moderne Straße. Planung, Bau und Verkehr vom 18. bis zum 20. Jahrhundert, edited by Hans-Liudger Dienel and Hans-Ulrich Schiedt, 49–76. Frankfurt am Main: Campus, 2010.

Nash, Roderick. *Wilderness and the American Mind*. New Haven: Yale University Press, 1967.

New Jersey Turnpike Authority. *Garden State Parkway*. Charleston, SC: Arcadia, 2013.

Nicholson, Geoff. *The Lost Art of Walking: The History, Science, Philosophy, and Literature of Pedestrianism*. New York: Riverhead, 2008.

Nolan, Mary. *The Transatlantic Century: Europe and America, 1890–2010*. Cambridge: Cambridge University Press, 2012.

Norton, Peter D. *Fighting Traffic: The Dawn of the Motor Age in the American City*. Cambridge, MA: MIT Press, 2008.

Nye, David E. *America as Second Creation: Technology and Narratives of New Beginnings*. Cambridge, MA: MIT Press, 2003.

———. *American Illuminations: Urban Lighting, 1800–1920*. Cambridge, MA: MIT Press, 2018.

———. *American Technological Sublime*. Cambridge, MA: MIT Press, 1994.

———. *America's Assembly Line*. Cambridge, MA: MIT Press, 2013.

———. *Electrifying America: Social Meanings of a New Technology, 1880–1940*. Cambridge, MA: MIT Press, 1990.

———. "Redefining the American Sublime, from Open Road to Interstate." In *Routes, Roads and Landscapes*, edited by Mari Hvattum, Janike Kampevold Larsen, Brita Brenna, and Beate Elvebakk, 99–112. Farnham: Ashgate, 2011.

O'Brien, William E. *Landscapes of Exclusion: State Parks and Jim Crow in the American South*. Amherst: University of Massachusetts Press, 2016.

O'Connell, Sean. *The Car and British Society. Class, Gender and Motoring 1896–1939*. Manchester: Manchester University Press, 1998.

Oettermann, Stephan. *The Panorama: History of a Mass Medium*. Translated by Deborah Lucas Schneider. New York: Zone Books, 1997.

Ogle, Vanessa. *The Global Transformation of Time 1870–1950*. Cambridge, MA: Harvard University Press, 2015.

Oldenziel, Ruth. *Making Technology Masculine: Men, Women and Modern Machines in America, 1870–1945*. Amsterdam: Amsterdam University Press, 1999.

Oreskes, Naomi, and John Krige, eds. *Science and Technology in the Global Cold War*. Cambridge, MA: MIT Press, 2014.

Otruba, Gustav. *A. Hitler's "Tausend-Mark-Sperre" und die Folgen für Österreichs Fremdenverkehr (1933–1938)*. Linz: Rudolf Trauner, 1983.

Packheiser, Christian. "Die Deutsche Alpenstraße als NS-Prestigeobjekt. Überlegungen zum Wechselverhältnis von Infrastruktur und Herrschaft im Nationalsozialismus." In *Die Deutsche Alpenstraße. Deutschlands älteste Ferienroute*, edited by Monika Kania-Schütz, 83–100. Munich: Volk, 2021.

Pagenstecher, Cord. "Die Automobilisierung des Blicks auf die Berge. Die Großglocknerstraße in Bildwerbung und Urlaubsalben." In *Tourisme et changements culturels. Tourismus und kultureller Wandel*, edited by Thomas Busset, Luigi Lorenzetti, and Jon Mathieu, 245–64. Zürich: Chronos, 2004.

———. *Der bundesdeutsche Tourismus. Ansätze zu einer Visual History: Urlaubsprospekte,*

Reiseführer, Fotoalben 1950–1990. Studien Zur Zeitgeschichte. vol. 4. Hamburg: Dr. Kovač, 2003.

Palmer, Scott W. *Dictatorship of the Air: Aviation Culture and the Fate of Modern Russia.* Cambridge: Cambridge University Press, 2006.

Parr, Joy. *Sensing Changes. Technologies, Environments, and the Everyday, 1953–2003.* Vancouver: UBC Press, 2010.

Patel, Kiran Klaus. *The New Deal: A Global History.* America in the World. Princeton: Princeton University Press, 2016.

———. *Soldiers of Labor: Labor Service in Nazi Germany and New Deal America, 1933–1945.* Publications of the German Historical Institute. New York: Cambridge University Press, 2005.

Petroski, Henry. *The Road Taken: The History and Future of America's Infrastructure.* New York: Bloomsbury, 2016.

Phillips, Sarah T. *This Land, This Nation: Conservation, Rural America and the New Deal.* New York: Cambridge University Press, 2007.

Pierce, Daniel S. *The Great Smokies: From Natural Habitat to National Park.* Knoxville: University of Tennessee Press, 2000.

———. "The Road to Nowhere: Tourism Development vs. Environmentalism in the Great Smoky Mountains." In *Southern Journeys. Tourism, History, and Culture in the Modern South,* edited by Richard D. Starnes, 196–214. Tuscaloosa: University of Alabama Press, 2003.

Pirie, Gordon. "British Air Shows in South Africa, 1932/33: 'Airmindedness,' Ambition and Anxiety." *Kronos* 35 (2009): 48–70.

Pomeroy, Earl S. *In Search of the Golden West: The Tourist in Western America.* New York: Knopf, 1957.

Pooley, Colin G., and Jean Turnbull. "Modal Choice and Modal Change: The Journey to Work in Britain since 1890." *Journal of Transport Geography* 8, no. 1 (2000): 11–24.

Pratt, Mary Louise. *Imperial Eyes: Travel Writing and Transculturation.* 2nd ed. London: Routledge, 2008.

Preston, Howard L. *Dirt Roads to Dixie: Accessibility and Modernization in the South, 1885–1935.* Knoxville: University of Tennessee Press, 1991.

Pritchard, Sara B. *Confluence. The Nature of Technology and the Remaking of the Rhône.* Cambridge, MA: Harvard University Press, 2011.

———. "Toward an Environmental History of Technology." In *The Oxford Handbook of Environmental History,* edited by Andrew C. Isenberg, 227–58. Oxford: Oxford University Press, 2014.

Pritchard, Sara B., and Carl A. Zimring, *Technology and the Environment in History.* Baltimore: Johns Hopkins University Press, 2020.

Probst, Robert. *Die NSDAP im Bayerischen Landtag 1924–1933.* Frankfurt am Main: Peter Lang, 1998.

Radde, Bruce. *The Merritt Parkway.* New Haven: Yale University Press, 1993.

Ramsey, E. Michele. "Driven from the Public Sphere: The Conflation of Women's Liberation and Driving in Advertising from 1910–1920." *Women's Studies in Communication* 29, no. 1 (2006): 88–112.

Reich, Justin. "Re-creating the Wilderness: Shaping Narratives and Landscapes in Shenandoah National Park." *Environmental History* 6 no. 2 (2001): 95–112.

Reinberger, Mark. "Architecture in Motion: The Gordon Strong Automobile Objective." https://franklloydwright.org/architecture-in-motion-the-gordon-strong-automobile -objective/, accessed May 25, 2021.

Reitsam, Charlotte. *Reichsautobahn-Landschaften im Spannungsfeld von Natur und Technik: Transatlantische und Interdisziplinäre Verflechtungen.* Saarbrücken: Verlag Dr. Müller, 2009.

Revill, George. *Railway.* London: Reaktion Books, 2012.

Rieger, Bernhard. "From People's Car to New Beetle: The Transatlantic Journeys of the Volkswagen Beetle." *Journal of American History* (2010): 91–115.

———. *The People's Car: A Global History of the Volkswagen Beetle.* Cambridge, MA: Harvard University Press, 2013.

Rigele, Georg. *Die Großglockner-Hochalpenstraße: Zur Geschichte eines österreichischen Monuments.* Vienna: WUV-Universitätsverlag, 1998.

———. "Sommeralpen-Winteralpen. Veränderungen im Alpinen durch Bergstraßen, Seilbahnen und Schilifte in Österreich." In *Umweltgeschichte. Zum historischen Verhältnis von Gesellschaft und Natur*, edited by Ernst Bruckmüller and Verena Winiwarter, 121–50. Vienna: öbv & hpt, 2000.

———. *Die Wiener Höhenstraße. Autos, Landschaft und Politik in den dreißiger Jahren.* Vienna: Turia & Kant, 1993.

Ritvo, Harriet. *The Dawn of Green: Manchester, Thirlmere, and Modern Environmentalism.* Chicago: University of Chicago Press, 2009.

Rodgers, Daniel T. *Atlantic Crossings: Social Politics in a Progressive Age.* Cambridge, MA: Harvard University Press, 1998.

Rogers, Jedediah S. *Roads in the Wilderness: Conflict in Canyon Country.* Salt Lake City: University of Utah Press, 2013.

Rome, Adam. *The Genius of Earth Day: How a 1970 Teach-In Unexpectedly Made the First Green Generation.* New York: Hill and Wang, 2013.

Rosa, Hartmut. "Social Acceleration: Ethical and Political Consequences of a Desynchronized High-Speed Society." *Constellations* 10, no. 1 (2003): 3–33.

Rose, Mark H., and Raymond A. Mohl. *Interstate: Highway Politics and Policy since 1939.* 3rd ed. Knoxville: University of Tennessee Press, 2012.

Rose, Mark H., and Bruce E. Seely. "Getting the Interstate System Built: Road Engineers and the Implementation of Public Policy, 1955–1985." *Journal of Policy History* 2, no. 1 (1990): 23–55.

Rose, Mark H., Bruce E. Seely, and Paul F. Barrett. *The Best Transportation System in the World: Railroads, Trucks, Airlines, and American Public Policy in the Twentieth Century.* Columbus: Ohio State University Press, 2006.

Rosenbaum, Adam T. *Bavarian Tourism and the Modern World, 1800–1950.* Publications of the German Historical Institute. New York: Cambridge University Press, 2016.

Rosenzweig, Roy, and Elizabeth Blackmar. *The Park and the People: A History of Central Park.* Ithaca, NY: Cornell University Press, 1992.

Roth, Ralf. *Das Jahrhundert der Eisenbahn. Die Herrschaft über Zeit und Raum 1814–1914.* Stuttgart: Thorbecke, 2005.

Roth, Ralf, and Colin Divall, eds. *From Rail to Road and Back Again? A Century of Transport Competition and Interdependency*. Ashgate: Farnham, 2015.

Rothman, Hal. *Devil's Bargains: Tourism in the Twentieth-Century American West*. Lawrence: University Press of Kansas, 1998.

Rugh, Susan Sessions. *Are We There Yet?: The Golden Age of American Family Vacations*. Cultureamerica. Lawrence: University Press of Kansas, 2008.

Runte, Alfred. *National Parks: The American Experience*. 4th ed. Lanham, MD: Taylor Trade, 2010.

———. *Trains of Discovery: Western Railroads and the National Parks*. rev. ed. Niwot, CO: Robert Rinehart, 1990.

Ruppmann, Reiner. *Schrittmacher des Autobahnzeitalters. Frankfurt und das Rhein-Main-Gebiet*. Schriften zur hessischen Wirtschafts- und Unternehmensgeschichte. vol. 10. Darmstadt: Hessisches Wirtschaftsarchiv, 2011.

Saraiva, Tiago. *Fascist Pigs: Technoscientific Organisms and the History of Fascism*. Cambridge, MA: MIT Press, 2016.

Sargeant, Jack, and Stephanie Watson. *Lost Highways: An Illustrated History of Road Movies*. London: Creation, 1999.

Schanetzky, Tim. *"Kanonen statt Butter": Wirtschaft und Konsum im Dritten Reich*. Munich: Beck, 2015.

Scharff, Virginia. *Taking the Wheel: Women and the Coming of the Motor Age*. New York: Free Press, 1991.

Schiedt, Hans-Ulrich. "Die Alpenstrassenfrage oder die 'prinzipielle Figur des Kreuzes.'" *Wege und Geschichte* no. 1 (2002): 34–39.

———. "Der Ausbau der Hauptstrassen in der ersten Hälfte des 20. Jahrhunderts." *Wege und Geschichte* no. 1 (2004): 12–25.

———. "Die Entwicklung der Strasseninfrastruktur in der Schweiz zwischen 1740 und 1910." *Jahrbuch für Wirtschaftsgeschichte* no. 1 (2007): 39–54.

Schipper, Frank. *Driving Europe: Building Europe on Roads in the Twentieth Century*. Technology and European History Series 3. Amsterdam: Aksant, 2008.

Schivelbusch, Wolfgang. *The Railway Journey. The Industrialization of Time and Space in the 19th Century*. Berkeley: University of California Press, 1986.

———. *Three New Deals: Reflections on Roosevelt's America, Mussolini's Italy, and Hitler's Germany, 1933–1939*. New York: Metropolitan Books, 2006.

Schlimm, Anette. *Ordnungen des Verkehrs: Arbeit an der Moderne. Deutsche und britische Verkehrsexpertise im 20. Jahrhundert*. Bielefeld: Transcript, 2011.

Schmoll, Friedemann. "Der Aussichtsturm. Zur Ritualisierung touristischen Sehens im 19. Jahrhundert." In *Reisebilder. Produktion und Reproduktion touristischer Wahrnehmung*, edited by Christoph Köck, 183–98. Münster: Waxmann, 2001.

Schmucki, Barbara. *Der Traum vom Verkehrsfluss. Städtische Verkehrsplanung seit 1945 im deutsch-deutschen Vergleich*. Frankfurt am Main: Campus, 2001.

Schöner, Helmut. *Berchtesgadener Fremdenverkehrs-Chronik 1923–1945*. Berchtesgaden: Fremdenverkehrsverband des Berchtesgadener Landes, 1974.

Schrag, Zachary M. "The Freeway Fight in Washington, DC: The Three Sisters Bridge in Three Administrations." *Journal of Urban History* 30, no. 5 (July 2004): 648–73.

Schueler, Judith. *Materialising Identity: The Co-construction of the Gotthard Railway and Swiss National Identity*. Amsterdam: Aksant, 2008.

Schultz, Jason. "To Render Inaccessible: The Sierra Club's Changing Attitude toward Roadbuilding." MA thesis, University of Maryland, 2008. http://hdl.handle.net/1903 /8984.

Schütz, Erhard, and Eckhard Gruber. *Mythos Reichsautobahn. Bau und Inszenierung der "Straßen des Führers" 1933–1941*. Berlin: Christoph Links, 1996.

Schuyler, David. *The New Urban Landscape: The Redefinition of City Form in Nineteenth-Century America*. Baltimore: Johns Hopkins University Press, 1986.

———. "The Sanctified Landscape: The Hudson River Valley, 1820 to 1850." In *Landscape in America*, edited by George F. Thompson, 93–109. Austin: University of Texas Press, 1995.

Schwarzenbach, Alexis. "Eine ungewöhnliche Erbschaft. Nutzung und Interpretation der Schlösser Ludwigs II. seit 1886." In*"Ein Bild von einem Mann": Ludwig II. von Bayern. Konstruktion und Rezeptions eines Mythos*, edited by Katharina Sykora, 27–47. Frankfurt am Main: Campus, 2004.

Seely, Bruce E. "Der amerikanische Blick auf die deutschen Autobahnen. Deutsche und amerikanische Autobahnbauer 1930–1965." *WerkstattGeschichte*, no. 21 (1998): 11–28.

———. *Building the American Highway System: Engineers as Policy Makers*. Philadelphia: Temple University Press, 1987.

———. "'Push' and 'Pull' Factors in Technology Transfer: Moving American-style Highway Engineering to Europe, 1945–1965." *Comparative Technology Transfer & Society* 2, no. 3 (2004): 229–46.

———. "The Scientific Mystique in Engineering: Highway Research in the Bureau of Public Roads, 1918–1940." *Technology and Culture* 25 (October 1984): 798–831.

———. "Visions of American Highways, 1900–1980." In *Geschichte der Zukunft des Verkehrs. Verkehrskonzepte von der Frühen Neuzeit bis zum 21. Jahrhundert*, edited by Hans-Liudger Dienel and Helmuth Trischler, 260–79. Frankfurt am Main: Campus, 1997.

Segal, Howard P. *Technological Utopianism in American Culture*. 20th anniversary ed. Syracuse, NY: Syracuse University Press, 2005.

Seidl, Alois. "Landwirtschaft (19./20. Jahrhundert)." *Historisches Lexikon Bayerns*. http:// www.historisches-lexikon-bayerns.de/Lexikon/Landwirtschaft (19./20. Jahrhundert), accessed May 24, 2018.

Seidler, Franz W. *Fritz Todt. Baumeister des Dritten Reiches*. Frankfurt am Main: Ullstein, 1988.

Seifert, Manfred. "Trachtenbewegung, Trachtenvereine." Historisches Lexikon Bayerns. http://www.historisches-lexikon-bayerns.de/Lexikon/Trachtenbewegung, Trachtenvereine, accessed April 17, 2018.

Seiler, Cotten. *Republic of Drivers: A Cultural History of Automobility in America*. Chicago: University of Chicago Press, 2008.

———. "'So That We as a Race Might Have Something Authentic to Travel By': African American Automobility and Cold-War Liberalism." *American Quarterly* 58 (2006): 1091–1117.

Sellars, Richard West. *Preserving Nature in the National Parks*. New Haven: Yale University Press, 1997.

Semmens, Kristin. *Seeing Hitler's Germany: Tourism in the Third Reich*. Basingstoke: Palgrave Macmillan, 2005.

Shaffer, Marguerite S. *See America First: Tourism and National Identity, 1880–1940*. Washington, DC: Smithsonian Institution Press, 2001.

Shankland, Robert. *Steve Mather of the National Parks*. New York: Knopf, 1951.

Shapiro, James. *Oberammergau: The Troubling Story of the World's Most Famous Passion Play*. New York: Pantheon, 2000.

Sheller, Mimi, and John Urry. "The New Mobilities Paradigm." *Environment and Planning A: Economy and Space* 38, no. 2 (2006): 207–26.

———. "Places to Play, Places in Play." In *Tourism Mobilities: Places to Play, Places in Play*, edited by Mimi Sheller and John Urry, 1–10. London: Routledge, 2004.

Siegelbaum, Lewis H. *Cars for Comrades: The Life of the Soviet Automobile*. Ithaca, NY: Cornell University Press, 2008.

———, ed. *The Socialist Car: Automobility in the Eastern Bloc*. Ithaca, NY: Cornell University Press, 2011.

Sies, Mary Corbin, and Christopher Silver, eds. *Planning the Twentieth-Century American City*. Baltimore: Johns Hopkins University Press, 1996.

Silver, Timothy. *Mount Mitchell and the Black Mountains: An Environmental History of the Highest Peaks in Eastern America*. Chapel Hill: University of North Carolina Press, 2003.

Skåden, Kristina. "The Map and the Territory: The Seventh International Road Congress, Germany 1934." *Transfers* 5, no. 1 (2015): 69–88.

Smith, David C. *City of Parks: The Story of Minneapolis Parks*. Minneapolis: Foundation for Minneapolis Parks, 2008.

Smith, Jason Scott. *Building New Deal Liberalism: The Political Economy of Public Works, 1933–1956*. New York: Cambridge University Press, 2006.

Solnit, Rebecca. *Wanderlust: A History of Walking*. New York: Penguin, 2000.

Sorin, Gretchen. *Driving While Black: African American Travel and the Road to Civil Rights*. New York: Liveright, 2020.

Sowards, Adam M. *The Environmental Justice: William O. Douglas and American Conservation*. Corvallis: Oregon State University Press, 2009.

Speck, Lawrence W. "Futurama." In *Norman Bel Geddes Designs America*, edited by Donald Albrecht, 288–315. New York: Abrams, 2012.

Speich, Daniel. "Mountains Made in Switzerland: Facts and Concerns in Nineteenth-Century Cartography." *Science in Context* 22, no. 3 (2009): 387–408.

———. "Wissenschaftlicher und touristischer Blick. Zur Geschichte der 'Aussicht' im 19. Jahrhundert." *Traverse* 3 (1999): 83–99.

Spence, Mark David. *Dispossessing the Wilderness: Indian Removal and the Making of the National Parks*. New York: Oxford University Press, 1999.

Spiro, Jonathan Peter. *Defending the Master Race: Conservation, Eugenics, and the Legacy of Madison Grant*. Burlington: University of Vermont Press, 2009.

Spode, Hasso. "Fordism, Mass Tourism and the Third Reich: The 'Strength through Joy' Seaside Resort as an Index Fossil." *Journal of Social History* 38 (2004): 127–55.

———. *Wie die Deutschen "Reiseweltmeister" wurden: Eine Einführung in die Tourismusgeschichte*. Erfurt: Landeszentrale für politische Bildung, 2003.

Starnes, Richard D. *Creating the Land of the Sky: Tourism and Society in Western North Carolina*. The Modern South. Tuscaloosa: University of Alabama Press, 2005.

Steininger, Benjamin. *Raum-Maschine Reichsautobahn: Zur Dynamik eines bekannt/unbekannten Bauwerks*. Berlin: Kadmos, 2005.

Stephan, Michael. "Feuchtwanger, Lion: Erfolg. Drei Jahre Geschichte einer Provinz, 1930." *Historisches Lexikon Bayerns*. http://www.historisches-lexikon-bayerns.de/artikel/artikel_44374, accessed July 11, 2013.

Stern, Alexandra Minna. *Eugenic Nation: Faults and Frontiers of Better Breeding in Modern America*. Berkeley: University of California Press, 2005.

Sternburg, Wilhelm von. *Ludwig Landmann: Ein Porträt*. Frankfurt am Main: S. Fischer, 2019.

Stilgoe, John. "Roads, Highways, and Ecosystems." http://nationalhumanitiescenter.org/tserve/nattrans/ntuseland/essays/roads.htm, accessed May 19, 2021.

Stilgoe, John R. *Metropolitan Corridor: Railroads and the American Scene*. New Haven: Yale University Press, 1983.

Sutter, Paul. *Driven Wild: How the Fight against Automobiles Launched the Modern Wilderness Movement*. Weyerhaeuser Environmental Books. Seattle: University of Washington Press, 2002.

———. "'A Retreat from Profit': Colonization, the Appalachian Trail, and the Social Roots of Benton MacKaye's Wilderness Advocacy." *Environmental History* 4, no. 4 (1999): 553–77.

Swart, Sandra, "Reviving Roadkill? Animals in the New Mobilities Studies." *Transfers* 5, no. 2 (2015): 81–101.

Swift, Earl. *The Big Roads: The Untold Story of the Engineers, Visionaries, and Trailblazers Who Created the American Superhighways*. Boston: Houghton Mifflin Harcourt, 2011.

Syon, Guillaume de. *Zeppelin! Germany and the Airship, 1900–1939*. Baltimore: Johns Hopkins University Press, 2002.

Tarr, Joel A. *The Search for the Ultimate Sink: Urban Pollution in Historical Perspective*. Akron, OH: University of Akron Press, 1996.

Taylor, Candacy. *Overground Railroad: The Green Book and the Roots of Black Travel in America*. New York: Abrams, 2020.

Tissot, Laurent. "How Did the British Conquer Switzerland? Guidebooks, Railways, Travel Agencies, 1850–1914." *Journal of Transport History* 16, no. 1 (March 1995): 21–54.

———. "Tourism in Austria and Switzerland: Models of Development and Crises, 1880–1960." In *Economic Crises and Restructuring in History. Experiences of Small Countries*, edited by Timo Myllyntaus, 285–302. St. Katharinen: Scripta Mercaturae, 1998.

Tooze, Adam. *The Wages of Destruction. The Making and Breaking of the Nazi Economy*. New York: Penguin, 2006.

Turner, James Morton. *The Promise of Wilderness: American Environmental Politics since 1964*. Seattle: University of Washington Press, 2013.

Uekötter, Frank. *The Green and the Brown: A History of Conservation in Nazi Germany*. Studies in Environment and History. Cambridge: Cambridge University Press, 2006.

Urry, John. *The Tourist Gaze: Leisure and Travel in Contemporary Societies*. London: Sage, 1990.

Utschneider, Ludwig. *Oberammergau im Dritten Reich*. 2nd ed. Oberammergau: Historischer Verein, 2012.

Vahrenkamp, Richard. *The German Autobahn, 1920–1945: Hafraba Visions and Mega Projects*. Lohmar: Josef Eul, 2010.

Vance Jr., James E. *The North American Railroad: Its Origin, Evolution, and Geography*. Baltimore: Johns Hopkins University Press, 1995.

Vanderbilt, Tom. *Traffic: Why We Drive the Way We Do (and What It Says about Us)*. New York: Knopf, 2008.

Vinsel, Lee. *Moving Violations: Automobiles, Experts, and Regulations in the United States*. Baltimore: Johns Hopkins University Press, 2019.

Waddy, Helena. *Oberammergau in the Nazi Era: The Fate of a Catholic Village in Hitler's Germany*. New York: Oxford University Press, 2010.

Walton, John K. *Histories of Tourism: Representation, Identity and Conflict*. Tourism and Cultural Change. Clevedon [England]: Channel View Publications, 2005.

Warner, Sam Bass. *Streetcar Suburbs: The Process of Growth in Boston, 1870–1900*. Publications of the Joint Center for Urban Studies. Cambridge, MA: Harvard University Press, 1962.

Weigold, Marilyn E. *Pioneering in Parks and Parkways. Westchester County, New York, 1895–1945*. Essays in Public Works History. vol. 9. Chicago: Public Works Historical Society, 1980.

Weingroff, Richard F. "From Names to Numbers: The Origins of the U.S. Numbered Highway System." https://www.fhwa.dot.gov/infrastructure/numbers.cfm, accessed July 24, 2017.

———. "The Lincoln Highway." https://www.fhwa.dot.gov/infrastructure/lincoln.cfm, accessed July 24, 2017.

Wells, Christopher. "The Changing Nature of Country Roads: Farmers, Reformers, and the Shifting Uses of Rural Space, 1880–1905." *Agricultural History* 80, no. 2 (2006): 143–66.

Wells, Christopher W. *Car Country: An Environmental History*. Seattle: University of Washington Press, 2012.

Wetterauer, Andrea. *Lust an der Distanz. Die Kunst der Autoreise in der "Frankfurter Zeitung."* Tübingen: Tübinger Verein für Volkskunde, 2007.

Whisnant, Anne Mitchell. "The Scenic Is Political: Creating Natural and Cultural Landscapes along America's Blue Ridge Parkway." In *The World beyond the Windshield. Roads and Landscapes in the United States and Europe*, edited by Christof Mauch and Thomas Zeller, 59–78. Athens: Ohio University Press, 2008.

———. *Super-Scenic Motorway: A Blue Ridge Parkway History*. Chapel Hill: University of North Carolina Press, 2006.

White, Richard. "'Are You an Environmentalist or Do You Work for a Living?' Work and Nature." In *Uncommon Ground: Rethinking the Human Place in Nature*, edited by William Cronon, 171–85. New York: Norton, 1995.

———. *The Organic Machine*. New York: Hill and Wang, 1995.

———. *Railroaded: The Transcontinentals and the Making of Modern America*. 1st ed. New York: Norton, 2011.

Wiater, Przemysław. "Droga Sudecka (Sudetenstraße) und Zakręt Śmierci (Todeskurve)." http://jbc.jelenia-gora.pl/Content/17942/de/seiten/1106.html, accessed January 5, 2022.

Williams, Michael. *Americans and Their Forests: A Historical Geography.* Studies in Environment and History. Cambridge: Cambridge University Press, 1989.

Wilson, Alexander. *The Culture of Nature: North American Landscape from Disney to the Exxon Valdez.* Cambridge, MA: Blackwell, 1992.

Winner, Langdon A. "Do Artifacts Have Politics?" *Daedalus* 109 (1980): 121–36.

Winter, James H. *Secure from Rash Assault: Sustaining the Victorian Environment.* Berkeley: University of California Press, 1999.

Wohl, Robert. *A Passion for Wings: Aviation and the Western Imagination, 1908–1918.* New Haven: Yale University Press, 1994.

———. *The Spectacle of Flight: Aviation and the Western Imagination, 1920–1950.* New Haven: Yale University Press, 2005.

Woolgar, Steve, and Geoff Cooper. "Do Artefacts Have Ambivalence? Moses' Bridges, Winner's Bridges and other Urban Legends in S&TS." *Social Studies of Science* 29 (1999): 433–49.

Woolner, David, and Henry Henderson, eds. *FDR and the Environment.* New York: Palgrave Macmillan, 2005.

Worster, Donald. *A Passion for Nature: The Life of John Muir.* Oxford: Oxford University Press, 2008.

Wrobel, Johannes. "Sudelfeld (SS-Berghaus und Hotel 'Alpenrose')." In *Der Ort des Terrors, Geschichte der nationalsozialistischen Konzentrationslager, vol. II: Frühe Lager, Dachau, Emslandlager,* edited by Wolfgang Benz and Barbara Distel, 505–7. Munich: Beck, 2005.

Young, Terence. "'A Contradiction in Democratic Government': W. J. Trent, Jr., and the Struggle to Desegregate National Park Campgrounds." *Environmental History* 14 (2009): 651–82.

Zapatka, Christian. "The American Parkways: Origins and Evolution of the Park-Road." *Lotus* 56 (1987): 97–128.

Zeller, Thomas. "Aiming for Control, Haunted by Its Failure: Towards an Envirotechnical Understanding of Infrastructures." *Global Environment* 10 (2017): 202–28.

———. *Driving Germany: The Landscape of the German Autobahn, 1930–1970.* Translated by Thomas Dunlap. New York: Berghahn, 2007.

———. "Loving the Automobile to Death? Injuries, Mortality, Fear, and Automobility in West Germany and the United States, 1950–1980." *Technikgeschichte* 86 (2019): 201–26.

———. "Loving Parks, Embracing Highways: Bernard DeVoto on Wilderness and Roads." In *Transatlantic Currents: A Tribute to David E. Nye's Contributions to American Studies,* edited by Anne Mørk, Kasper Grotle Rasmussen, and Jørn Brøndal, 123–33. Heidelberg: Universitätsverlag Winter, 2021.

———. "Mein Feind, der Baum: Verkehrssicherheit, Unfalltote, Landschaft und Technik in der frühen Bundesrepublik." In *Mit dem Wandel leben. Neuorientierung und Tradition in der Bundesrepublik der 1950er und 60er Jahre,* edited by Friedrich Kießling and Bernhard Rieger, 247–66. Cologne: Böhlau, 2010.

———. "Molding the Landscape of Nazi Environmentalism: Alwin Seifert and the Third Reich." In *How Green Were the Nazis? Nature, Environment, and Nation in the Third Reich,* edited by Franz Josef Brüggemeier, Mark Cioc, and Thomas Zeller, 147–70. Athens: Ohio University Press, 2005.

———. "Staging the Driving Experience: Parkways in Germany and the United States." In *Routes, Roads and Landscapes*, edited by Mari Hvattum, Janike Kampevold Larsen, Brita Brenna, and Beate Elvebakk, 125–38. Farnham: Ashgate, 2011.

———. *Straße, Bahn, Panorama. Verkehrswege und Landschaftsveränderung in Deutschland 1930 bis 1990*. Frankfurt am Main: Campus, 2002.

Zimring, Carl. "'Neon, Junk, and Ruined Landscape': Competing Visions of America's Roadsides and the Highway Beautification Act of 1965." In *The World beyond the Windshield: Roads and Landscapes in the United States and Europe*, edited by Christof Mauch and Thomas Zeller, 94–107. Athens: Ohio University Press, 2008.

Zochert, Donald, ed. *Walking in America*. New York: Knopf, 1974.

Zschokke, Walter. *Die Strasse in der vergessenen Landschaft. Der Sustenpass*. Zurich: gta, 1997.

Index

Page numbers in *italics* indicate maps or photographs.